T0314052

Krylov Subspace Methods with Application in
Incompressible Fluid Flow Solvers

Krylov Subspace Methods with Application in Incompressible Fluid Flow Solvers

Iman Farahbakhsh
Amirkabir University of Technology
Tehran, Iran

Registered Offices
John Wiley & Sons, Inc., 111 River Street, Hoboken, NJ 07030, USA
John Wiley & Sons Ltd, The Atrium, Southern Gate, Chichester, West Sussex, PO19 8SQ, UK

Editorial Office
The Atrium, Southern Gate, Chichester, West Sussex, PO19 8SQ, UK

For details of our global editorial offices, customer services, and more information about Wiley products visit us at www.wiley.com.

Wiley also publishes its books in a variety of electronic formats and by print-on-demand. Some content that appears in standard print versions of this book may not be available in other formats.

Library of Congress Cataloging-in-Publication Data

Names: Farahbakhsh, Iman, 1984- author.
Title: Krylov subspace methods with application in incompressible fluid
   flow solvers / Iman Farahbakhsh.
Description: Hoboken, NJ : Wiley, 2020. | Includes bibliographical
   references and index.
Identifiers: LCCN 2020014447 (print) | LCCN 2020014448 (ebook) | ISBN
   9781119618683 (hardback) | ISBN 9781119618690 (adobe pdf) | ISBN
   9781119618706 (epub)
Subjects: LCSH: Sparse matrices. | Fluid dynamics–Mathematical models. | Computational fluid
   solutions. | Navier-Stokes equations–Numerical
   dynamics. | Numerical analysis.
Classification: LCC QA188 .F36 2020 (print) | LCC QA188 (ebook) | DDC
   512.9/434–dc23
LC record available at https://lccn.loc.gov/2020014447
LC ebook record available at https://lccn.loc.gov/2020014448

Cover Design: Wiley
Cover Image: © anuskiserrano/Shutterstock

Set in 9.5/12.5pt STIXTwoText by SPi Global, Chennai, India

Printed and bound by CPI Group (UK) Ltd, Croydon, CR0 4YY

10 9 8 7 6 5 4 3 2 1

In memory of the great soul of my dear mother,
Fatima Bakhshayesh,
Dedicated to my dear wife, Dr. Zahra Kalantari

Contents

List of Figures

List of Tables

Preface

I remember that in 2009, when I was trying to apply the Krylov subspace methods to incompressible fluid flow solvers codes, in primary and secondary formulations of Navier-Stokes equations, I felt a lack of resources that would link fundamental issues of the numerical linear algebra to the computational fluid dynamics. Following on from the development of the Krylov subspace method codes that I used to solve the equation systems derived from the elliptic part of the governing equations of the fluid dynamics, I compiled lecture notes to use as supplemental topics in advanced numerical computations and computational fluid dynamics courses. In almost all computational fluid dynamics textbooks, only a few stationary solvers of elliptic equations have been mentioned, and methods based on orthogonalization have been neglected. On the other hand, orthogonalization based methods, and in particular Krylov subspace methods, need to be basically articulated, that may not fit into a textbook exclusively devoted to computational fluid dynamics. So I decided to expand the lecture notes in LATEX and adjust them so that, along with the algorithm of the methods, the incompressible flow solver codes that used the developed subroutines of the Krylov subspace methods were also included. These lecture notes were completed in 2018 and were edited in the following two years to form this book.

The structure of the book is formed in a way that covers the basic topics about the common Krylov subspace methods and their preconditioned versions which is accompanied by the selected topics in storage formats of sparse matrices. An important part of the book is devoted to the numerical analysis of Krylov subspace methods, and we explain how they play a role in an incompressible flow solver.

I have tried to organize the contents of the book in such a way that it can be introduced as a textbook for graduate students in courses such as advanced numerical computations, computational fluid dynamics, advanced numerical analysis, numerical linear algebra and numerical solution of PDEs. Subroutines and main programs of the various sections of the book, along with descriptions, are included in the Appendices, which engages the reader in developing and applying Krylov subspace solvers on various issues, and especially computational fluid dynamics. The code included in the book appendices are also available in the companion website.

The book chapters end with exercises which in some cases can be addressed as course projects. In many exercises, the reader is directed to use the appendices subroutines to solve problems and, in some cases, to develop them. In the exercises, the most important challenges in optimizing the solution of the linear system of equations using the Krylov subspace methods are addressed and the reader is encouraged to build his own experience in solving them using the appendices codes. I hope that the contents of this book satisfy the readers' expectations and I enthusiastically look forward to receive the comments and reviews of the respected readers. Obviously, future editions will come from effective feedback from my dear readers and colleagues.

January 7, 2020 *Iman Farahbakhsh*
Tehran

About the Companion Website

The companion website for this book is at:

www.wiley.com\go\Farahbakhs\KrylovSubspaceMethods

The website includes:

- Code of Appendices

Scan this QR code to visit the companion website.

1

Introduction

1.1 Motivation

In order to gain a deep and accurate understanding of physical phenomena, humanity has always been looking for a way to mathematically model these phenomena. With the increase of human knowledge and understanding of the physical phenomenology of natural events, mathematical modeling has gradually become more complete and complex. The purpose of mathematical modeling of a physical phenomenon is the expression of the dependence of physical quantities, either vector or scalar, on the components of Euclidean space and time. In general, we can say that almost all of the phenomena in the universe, even the events in which humans are involved, can be modeled using partial differential equations (PDE). It should be noted that finding an explicit solution for these equations is often cumbersome, if not impossible.

Incompressible fluid flow, as a real representation of a physical phenomenon, can be modeled using PDEs. The basis of computational fluid dynamics (CFD) is the basic governing equations of fluid dynamics. These equations provide a mathematical model for describing the behavior of Newtonian fluids. These equations are known as the Navier–Stokes equations, which were first proposed by a French engineer Claude-Louis Navier in 1822 based on a molecular model. It is interesting that the scientific committee of that time found these equations inappropriate, from the physical point of view, for many materials, especially fluids. After 23 years, in 1845, a British scientist, Sir George Gabriel Stokes, derived similar equations using the continuum theory. In this section, we intend to briefly present these equations and introduce different methods of solution.

Krylov Subspace Methods with Application in Incompressible Fluid Flow Solvers, First Edition.
Iman Farahbakhsh.
© 2020 John Wiley & Sons Ltd. Published 2020 by John Wiley & Sons Ltd.
Companion Website: www.wiley.com/go/Farahbakhs/KrylovSubspaceMethods

1.1.1 Governing Equations

The Navier–Stokes equations for incompressible fluid motion as governing equations, are the mathematical expression of three physical laws of mass, momentum and energy conservation which can be expressed as

$$\frac{\partial u_j}{\partial x_j} = 0, \tag{1.1}$$

$$\frac{\partial u_i}{\partial t} + u_j \frac{\partial u_i}{\partial x_j} + \frac{1}{\rho} \frac{\partial p}{\partial x_i} = v\Delta u_i + f_i, \tag{1.2}$$

$$\frac{\partial e}{\partial t} + u_j \frac{\partial e}{\partial x_j} = \mathcal{E} + \frac{1}{\rho} \frac{\partial}{\partial x_j} \left(\kappa \frac{\partial T}{\partial x_j} \right), \tag{1.3}$$

where, u_j is the velocity component, p is the pressure, x_j is the coordinate component, ρ is the density, f_i is the component of external force per unit mass, Δ is the Laplacian operator, e is the internal energy per unit mass, v is the kinematic viscosity, κ is the coefficient of thermal conductivity, T is the temperature, and $\mathcal{E}$ is the rate of dissipation of mechanical energy per unit mass or more often called the viscous dissipation function per unit mass that can be expressed as

$$\mathcal{E} = \frac{v}{2} \left(\frac{\partial u_i}{\partial x_j} + \frac{\partial u_j}{\partial x_i} \right) \left(\frac{\partial u_i}{\partial x_j} + \frac{\partial u_j}{\partial x_i} \right). \tag{1.4}$$

In Equation (1.1) to Equation (1.3), it is assumed that the fluid is incompressible and homogeneous, that is, density is constant in time and space. Based on the internal energy definition and considering the constant conduction heat transfer coefficient and specific heat coefficient, the energy equation, Equation (1.3), can be written as

$$\frac{\partial T}{\partial t} + u_j \frac{\partial T}{\partial x_j} = \frac{\mathcal{E}}{c_p} + \frac{\kappa}{\rho c_p} \frac{\partial^2 T}{\partial x_j^2}, \tag{1.5}$$

where $\frac{\kappa}{\rho c_p}$ is the thermal diffusivity and is usually given as α in the texts. Equation (1.1), Equation (1.2), and Equation (1.5) are complex sets of elliptic and parabolic equations, and the unknowns are velocity, pressure, and temperature. Since the temperature explicitly appears in the energy equation, it can be distinguished from the two other equations. However, the fluid properties, c_p, κ, and v are usually temperature dependent. Therefore, the separation of the energy equation from the two other equations is only true when these properties, especially the viscosity, are independent of temperature. In many cases, the temperature variations are negligible and can be neglected, and one can ignore the solution of the energy equation. With these interpretations, if the velocity field is specified, obtaining the temperature distribution is simple.

1.1.2 Methods for Solving Flow Equations

Rarely, we can find a closed form for analytical solution of Equation (1.1), Equation (1.2) and Equation (1.5) with suitable boundary and initial conditions. Therefore, we need to look for a numerical method to solve these equations. First, the desired equations and the boundary and initial conditions must be discretized in space and time. Although there are very different methods for discretizing and solving the equations, eventually spatial and temporal discretization leads to solving the following system of equations, that is

$$Ax^{n+1} = b^{n+1}. \tag{1.6}$$

Here, the matrix of coefficients A and vector b is specified and x is the unknown vector. The ultimate goal of this book is to provide efficient methods for solving the system of equations and that will be addressed in detail in the following chapters. This book explores the mathematical foundation and numerical analysis of Krylov[1] subspace methods in dealing with the system of equations. In this book, some methods for solving the incompressible fluid flow are considered and Krylov subspace methods will be used as the solver of the elliptic part of equations.

Several methods have been developed to solve the governing equations of the incompressible viscous fluid flow. These methods can be divided into two main categories, methods used to calculate the primitive variables such as velocity and pressure, and methods used to calculate the derived variables such as vorticity and stream function. Figure 1.1 shows some of the existing numerical methods for solution of the fluid dynamics governing equations. In this book, we explain some of the mentioned approaches in Figure 1.1 and examine the efficiency of the Krylov subspace methods in solution of the elliptic parts of equations.

---

1 Aleksey Nikolaevich Krylov (1863–1945) was a Russian naval engineer, applied mathematician, and memoirist. In 1931 he published a paper on what is now called the Krylov subspace and Krylov subspace methods. The paper deals with eigenvalue problems, namely, with computation of the characteristic polynomial coefficients of a given matrix. Krylov was concerned with efficient computations and, as a computational scientist, he counts the work as a number of separate numerical multiplications; something not very typical for a 1931 mathematical paper. Krylov begins with a careful comparison of the existing methods that include the worst-case-scenario estimate of the computational work in the Jacobi method. Later, he presents his own method which is superior to the known methods of that time and is still widely used.

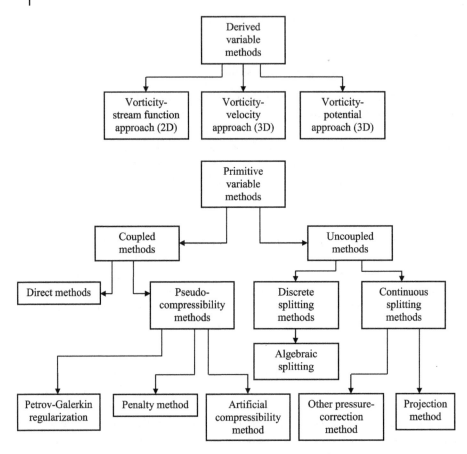

Figure 1.1 Some numerical methods for solving the fluid dynamics governing equations.

1.2 History of Krylov Subspase Methods

Methods based on the orthogonalization were developed in the early 1950s. Lanczos (1950) proposed a method based on two orthogonal vector sequences. His main motivation in the development of this method was the result of eigenvalues problems. The most prominent feature of his method was to simplify the original matrix into a tridiagonal form. Lanczos (1952) later applied his method to solve symmetric linear systems. An important property for the convergence of this method in solving linear systems is that the approximate solution vector can be expressed by multiplying the initial residual vector in a polynomial of the coefficients matrix. Hestenes and Stiefel (1952) developed a method that was a classical definition of the conjugate gradient (CG) method for solving linear systems. In fact, this method is a simplified form of the Lanczos method for symmetric matrices, in other words, two orthogonal sequences are reduced to an orthogonal sequence. In the Lanczos method, the three-term recurrences are used and the two-term recursive relations are used in the Hestenes–Stiefel method. By combining two two-term recursive relations, by eliminating vectors of search direction, we can extract the Lanczos method (Barrett et al., 1994).

Arnoldi (1951) developed a new method with combined features of Hestenes–Stiefel and Lanczos methods. This method, like the Lanczos method, can be used for asymmetric matrices and does not use vectors of search direction. Reid (1971) noted in an article that the most important field for operation of the CG method is sparse definite systems, which led to the CG method being considered again. Fletcher (1976) presented a special type of Lanczos approach that was similar to the CG method. This method, called the bi-conjugate gradient (BiCG), is suitable for solving asymmetric systems and has two two-term recursive relations.

The development of Krylov subspace methods for solving asymmetric linear systems is an active research field and the development of new methods is still ongoing. Here, only the most popular and most used methods will be addressed, especially the methods in which there are extensive computing experiences. In an article presented by Freund et al. (1992), the methods developed up to 1991 were examined. Also, in the papers presented by Yang (1992) and Tong (1992), a large numerical comparison was made between different methods.

Following the development of the generalized minimal residual (GMRES) method by Saad and Schultz (1986), several suggestions were made to reduce the required memory and computational cost of this method. The most obvious solution to this problem is to use the reinitiation algorithm of GMRES(m). Another approach is to restrict the GMRES search to a suitable subspace of some higher-dimensional Krylov subspace. All methods that are based on this idea are called preconditioned GMRES methods. The simplest of these methods uses a fixed polynomial as a preconditioner (Johnson et al., 1983; Saad, 1985; Nachtigal et al., 1990). In more complex methods, a preconditioner polynomial may be adapted to different iterations (Saad, 1993) or essentially other iteration based methods are used as preconditioner (Vorst and Vuik, 1991; Axelsson and Vassilevski, 1991).

Recent studies have focused on improving BiCG with more favorable characteristics. These features include preventing breakdown, avoiding the use of matrix transpose, optimizing the use of matrix–vector products, smooth convergence, and exploiting the operations performed to form the Krylov subspace with A^T for further reduction of the residual. In BiCG, the use of matrix transpose in matrix–vector multiplications can be problematic. In fact, by increasing the degree of Krylov subspace in each iteration, the number of matrix–vector multiplications in the BiCG method doubles as compared with the CG method.

Hence, several methods have been proposed to overcome this problem. The most interesting of these methods, the conjugate gradient squared (CGS), was introduced by Sonneveld (1989). In this method, the BiCG polynomial square is computed without the need for a matrix transpose, so the need for matrix transpose is also eliminated. When the BiCG method converges, in most cases the CGS method can also be considered as another option that has faster convergence. However, the CGS method has been characterized by breakdown and irregular convergence of the BiCG method and often exacerbates it (Van der Vorst, 1992). The CGS method paves the way for developing methods that are generally based on the multiplication of polynomials. In these methods, the BiCG polynomial is multiplied in a polynomial of the same degree and the necessity of using the matrix transpose is eliminated. Bi-conjugate gradient stabilized (BiCGSTAB) is an example of this group of methods proposed by Van der Vorst (1992). In this method, by means of local minimization, a polynomial is defined which helps to smooth the convergence process. Gutknecht noted that the BiCGSTAB method could be considered as a product of the BiCG and GMRES(1) methods and he proposed the combination of the BiCG method with GMRES(2) for the even numbered iteration steps (Gutknecht, 1993). It was expected that this idea would lead to better convergence in cases where the coefficients matrix has complex eigenvalues. A different form of this method, which has more accuracy and efficiency, was presented in a paper by Sleijpen and Fokkema (1993), in which the method of combining the BiCG method and GMRES(m), for relatively low values of m, was presented.

In addition to the BiCGSTAB method, other methods are also proposed to smooth the convergence of the CGS method. One of the ideas that arises is the use of the quasi-minimal residual (QMR) to obtain a smoother convergence. Freund proposed the QMR version of the CGS method and called it TFQMR (Freund, 1993). Numerical experiments show that the TFQMR has the desired characteristics of the CGS and corrects its irregular convergence. The low computational cost, the smooth convergence process, and the lack of need for the transpose of the matrix makes TFQMR a suitable alternative to CGS. Chan et al. (1994) extracted the QMR version of the BiCGSTAB method and called it QMRCGSTAB. These methods have a much smoother convergence process than CGS and BiCGSTAB, with less computational cost.

Certainly, it is not possible to comment on the superiority of a particular method. Each of the Krylov subspace methods can overcome a set of specific problems and the major issue is identifying these problems and developing new methods to solve them. For each group of Krylov subspace methods, there are systems that some of the methods are able to solve, and others are either encountering problems or basically unable to solve them. This is illustrated in an article by Nachtigal et al. (1992).

Iterative methods never reach the precision of direct methods and cannot overcome them. For some problems, the use of iterative methods is more attractive, and for others, direct methods. Direct methods for solving large linear systems impose a large amount of computational cost in comparison with iterative methods. Finding appropriate methods and preconditioners for a group of systems that have not been solved yet due to CPU constraints, memory, convergence issues, ill-conditioned matrices, etc., is one of the biggest goals that has been taken into account in this research field.

1.3 Scope of Book

In this book, we try to explain the fundamentals of Krylov subspace solver theory as much as possible and examine their capabilities and deficiencies in dealing with various problems. The solvers provided in the appendices can be used as a software package in all CFD codes. This package consists of 15 Krylov subspace solvers, which can be selected depending on the type of problem. On the whole, these solvers can solve a variety of linear sparse systems, which will be discussed in detail in Chapter 4.

1.3.1 The General Structure of Solver

Figure 1.2 shows the general process of solving the linear system of equations by the code developed based on the Krylov subspace methods. The package of developed solvers is able to integrate with an algebraic multigrid solver, which has been developed as an independent research topic, and displays excellent performance. In other words, the multigrid solver can be used as a preconditioner and an error smoother.

The developed package is able to find its input, the system of equations derived from a particular problem and generates the solution vector. To constitute a set of equations as an input to this package, one can also use the prefabricated matrices of Matrix Market and Harwell-Boeing templates; see Section 2.4. In this book we use subroutines that are able to produce right-hand side vectors and sparse matrices in a compact format by discretizing equations on structured and semistructured grids which provides the required inputs for solving the system of equations. The solver can also get the compact format of sparse matrices derived from unstructured grids as an input. In Figure 1.3, the content of the solver package is shown.

In this book we introduce eight compact formats of sparse matrices which are widely used in the matrix–vector product that is an essential part of Krylov subspace solvers algorithm. In Chapter 4, the effect of using the various compact

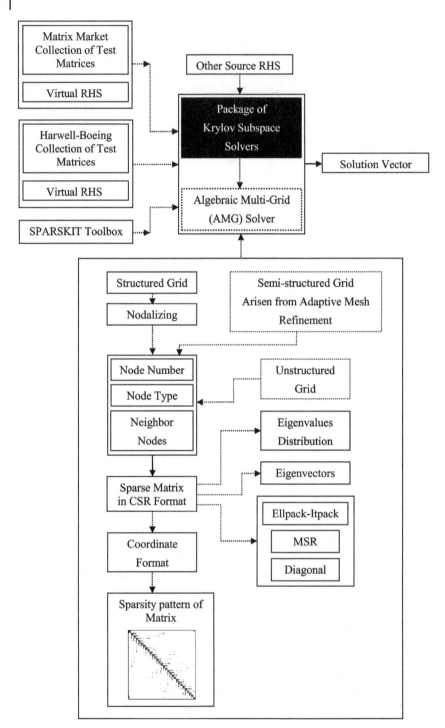

Figure 1.2 General process of the linear set of equations using the Krylov subspace based package.

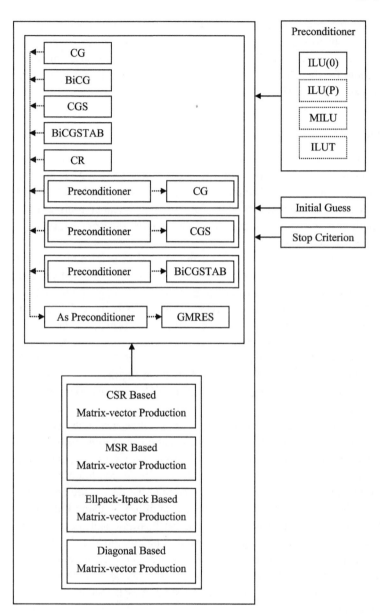

Figure 1.3 The content of the solver package.

formats on the CPU-time will be assayed. In the developed solvers, preconditioning by incomplete factorization can be used, in this book we merely examine the effect of the preconditioner ILU(0). Using some of the Krylov subspace methods as preconditioner for the GMRES method and evaluating their performance will be also addressed in this book.

1.3.2   Review of Book Content

At the beginning of Chapter 2, the partial differential equations, elliptic operators and advection–diffusion equation will be briefly described. Section 2.3 is devoted to finite difference method and discretizing the Laplace operator. In Section 2.4 we will introduce the sparse matrices which constitute the linear systems for evaluating the methods in Chapter 4. At the end of this chapter, the storage formats of sparse matrices will be discussed.

In Chapter 3, the underlying foundations and general principles for deriving some of the Krylov subspace methods, as the basis of this book, will be discussed. At the beginning of this chapter, the general principles of the projection method are described as the cornerstone of the Krylov subspace methods, and in Section 3.3 we will define the Krylov subspace. In Section 3.4, principles, derivation approaches and convergence theory of CG method are studied. The remainder of the chapter deals with investigation of minimal residual (MINRES), GMRES, conjugate residual (CR), and BiCG methods. Finally, in the Section 3.9, we will explain the principles of extracting CGS and BiCGSTAB methods, as a subset of the transpose-free methods.

Chapter 4 is devoted to numerical analysis of Krylov subspace methods. At the beginning of this chapter, we will examine the capabilities of different methods in solving equations with symmetric positive definite, symmetric indefinite, and asymmetric matrices of coefficients. In Section 4.2, the preconditioning of the system of equations is discussed and, as an example, we derive the algorithm of the preconditioned CG method. Then, the preconditioning using the ILU(0) factorization is described and further we will address the numerical analysis of the solution of the equation system using the preconditioned methods. Subsequently, the numerical analysis of the combination of GMRES method and other methods is presented, and then we will examine the effect of the storage formats of sparse matrices on the CPU-time. The solution of the singular systems and problems with the pure Neumann boundary conditions is one of the issues that will be addressed in the final section of Chapter 4.

Chapter 5 is devoted primarily to the theoretical basics and analysis of the projection method. This chapter will deal with the general framework, discretization and implementation of the projection method. In the solution process of the projection method one can find the pseudo-pressure Poisson's equation with pure

Neumann boundary conditions which should be solved via an optimum solver. The author tries to give an accurate, detailed description of the projection method as a primary variable scheme, along with its characteristics, according to the educational goal. The numerical experiences of Sections 4.5.1 and 4.5.2 will show that the Krylov subspace solvers in dealing with the singular systems have a remarkable performance with respect to stationary solvers. The author explains, based on the supportive materials of Chapter 4, where the Krylov subspace methods should be applied in the solution process of the projection method. At the end of this chapter, some numerical experiments such as the problem of vortex shedding around the circular and four leaf cylinder and oscillating circular cylinder in a quiescent fluid are investigated.

2

Discretization of Partial Differential Equations and Formation of Sparse Matrices

2.1 Introduction

The most common way to solve partial differential equations is to discretize them, to estimate them with equations with a finite number of unknowns. The resulting system of equations from this discretization usually have large and sparse matrices; in fact, these matrices have a small number of non-zero elements. There are different methods for the discretization of a partial differential equation. The simplest method is central difference estimation for partial differential operators. In this chapter, we first discuss briefly the partial differential equations and then introduce the finite difference method. In the following, a definition of sparse matrices is presented and some of the benchmark matrices that are used in the next chapters for comparing the solvers are introduced. Finally, the storage formats of sparse matrices will be addressed.

2.2 Partial Differential Equations

Physical phenomena are typically modeled by equations that combine partial derivatives of physical quantities such as force, momentum, velocity, energy, temperature, and so on. Analytical solutions rarely exist for these equations and numerical approaches are used in general. In this chapter, we address only those partial differential equations that are used in incompressible flow problems. Dimensions of problems in this chapter do not exceed two, where x is considered a spatial variable for one dimensional problems while x_1 and x_2 are used for two dimensional problems.

Krylov Subspace Methods with Application in Incompressible Fluid Flow Solvers, First Edition.
Iman Farahbakhsh.
© 2020 John Wiley & Sons Ltd. Published 2020 by John Wiley & Sons Ltd.
Companion Website: www.wiley.com/go/Farahbakhs/KrylovSubspaceMethods

2.2.1 Elliptic Operators

One of the common partial differential equations which describe many physical phenomena is Poisson's equation, i.e.,

$$\frac{\partial^2 u}{\partial x_1^2} + \frac{\partial^2 u}{\partial x_2^2} = f(x_1, x_2), \quad x = \begin{pmatrix} x_1 \\ x_2 \end{pmatrix} \tag{2.1}$$

Here, x_1 and x_2 are spatial variables in the domain of $\Omega \in \mathbb{R}^2$. The equation holds only for interior points of the domain while boundary points usually meet in one of these boundary conditions

$$u(x) = \phi(x) \quad \text{Dirichlet boundary condition}$$

$$\frac{\partial u}{\partial n}(x) = \psi(x) \quad \text{Neumann boundary condition}$$

$$\frac{\partial u}{\partial n}(x) + \alpha(x)u(x) = \gamma(x) \quad \text{Cauchy boundary condition.}$$

where n is a unit vector normal to the boundary and pointing outwards. It should be also noted that the Neumann boundary condition is a special case of the Cauchy boundary condition with $\alpha = 0$. For a unit vector v (with components v_1 and v_2) the directional derivative $\frac{\partial u}{\partial n}$ can be written as

$$\begin{aligned} \frac{\partial u}{\partial n}(x) &= \lim_{h \to 0} \frac{u(x+hv)-u(x)}{h} \\ &= \frac{\partial u}{\partial x_1}v_1 + \frac{\partial u}{\partial x_2}v_2 \\ &= \nabla u.v \end{aligned} \tag{2.2}$$

where ∇ is the gradient operator which gives the vector

$$\nabla u = \begin{pmatrix} \frac{\partial u}{\partial x_1} \\ \frac{\partial u}{\partial x_2} \end{pmatrix}. \tag{2.3}$$

In fact, Poisson's equation often represents the asymptotic kind of a time dependent problem. For instance, Poisson's equation describes the steady state distribution of temperature in the domain of Ω in the presence of the time independent heat source f. The boundary conditions must be such that the heat flux is well modeled on the boundaries. In an special case when $f = 0$, that means

$$\Delta u = 0,$$

a special type of Poisson's equation, called Laplace's equation, is derived which its solution is a harmonic function.

In problems with pure Neumann boundary conditions, it is of great importance to note that there is no unique solution for Laplace's equation such that if u is an answer to a problem, then for any constant value c, $u + c$ is also a solution for the problem. In Chapter 4, the numerical solution of such problems will be explored extensively. Here, Δ is Laplace's operator which appears in many physical and mechanical models and is defined as

$$\Delta = \frac{\partial^2}{\partial x_1^2} + \frac{\partial^2}{\partial x_2^2}.$$

These models often lead to more general elliptic operators such as

$$L = \frac{\partial}{\partial x_1}\left(a\frac{\partial}{\partial x_1}\right) + \frac{\partial}{\partial x_2}\left(a\frac{\partial}{\partial x_2}\right) = \nabla.(a\nabla), \tag{2.4}$$

where, the scalar function a depends on the coordinates and may represent certain physical properties such as density, porosity, and so on. Here, we define some of the operators briefly. The operator ∇ can be considered as a vector, including the components $\frac{\partial}{\partial x_1}$ and $\frac{\partial}{\partial x_2}$. When this operator is applied to a scalar function, it is called a gradient and its product is a vector. The inner product of this operator and vector v results in a scalar that represents divergence of the vector

$$\text{div } v = \nabla.v = \frac{\partial v_1}{\partial x_1} + \frac{\partial v_2}{\partial x_2}.$$

Applying the divergence operator to $u = a\nabla$, where a is a scalar function, results in the operator L, as mentioned in Equation (2.4).

Operator

$$L = \begin{pmatrix} \frac{\partial}{\partial x_1}(a_1\frac{\partial}{\partial x_1} + a_2\frac{\partial}{\partial x_2}) \\ \frac{\partial}{\partial x_2}(a_1\frac{\partial}{\partial x_1} + a_2\frac{\partial}{\partial x_2}) \end{pmatrix} = \nabla(a.\nabla)$$

is a generalized form of Laplace's operator and is used for mathematical modeling of some physical phenomena in anisotropic and inhomogeneous domains. The coefficients a_1 and a_2 are functions of coordinate variables and express the directional and spatial dependency of physical properties such as porosity (in fluid flow) or conduction heat transfer coefficient (in heat transfer). In fact, the above operator can be considered as $L = \nabla.(A\nabla)$, where A is a 2×2 matrix and acts on two components of ∇.

2.2.2 Convection–Diffusion Equation

Many physical problems are associated with a combination of convection and diffusion phenomena. Therefore, they can be described by a convection–diffusion equation with this formulation

$$\frac{\partial u}{\partial t} + b_1\frac{\partial u}{\partial x_1} + b_2\frac{\partial u}{\partial x_2} = \nabla.(a\nabla)u + f,$$

or

$$\frac{\partial u}{\partial t} + b.\nabla u = \nabla.(a\nabla)u + f,$$

which in steady state form is simplified as

$$\boldsymbol{b}.\nabla u - \nabla.(a\nabla)u = f. \tag{2.5}$$

It should be noted that in some cases the vector $\boldsymbol{b}$ is very large, which may lead to problems in the discretization of the equation or the iteration based methods of solution.

2.3 Finite Difference Method

The finite difference method is based on estimates of derivatives obtained by Taylor series expansion in a partial differential equation. This method is simple and easily applicable such that its implementation on problems with simple domains and uniform grids is highly efficient. The matrices resulting from this discretization approach often have a simple pattern.

If a second order central difference approximation is applied for discretization of $\frac{\partial^2}{\partial x_1^2}$ and $\frac{\partial^2}{\partial x_2^2}$ in Laplace's operator, and size of grid in directions x_1 and x_2 are h_1 and h_2 respectively, the following approximation holds

$$\Delta u(\boldsymbol{x}) \approx \frac{u(x_1 + h_1, x_2) - 2u(x_1, x_2) + u(x_1 - h_1, x_2)}{h_1^2}$$
$$+ \frac{u(x_1, x_2 + h_2) - 2u(x_1, x_2) + u(x_1, x_2 - h_2)}{h_2^2}.$$

In a special case when $h_1 = h_2$, the approximation is simplified and written as (McDonough, 2008)

$$\Delta u(\boldsymbol{x}) \approx \frac{1}{h^2}[u(x_1 + h, x_2) + u(x_1 - h, x_2) + u(x_1, x_2 + h) \tag{2.6}$$
$$+ u(x_1, x_2 - h) - 4u(x_1, x_2)].$$

This is called the five-point stencil of Laplace's operator. The stencil of this approximation is shown in Figure 2.1. Another form of this stencil can be assumed by four points $u(x_1 \pm h, x_2 \pm h)$ with a similar formulation of Equation (2.6), but the grid size will change; see Figure 2.1. Finite difference approximation in Equation (2.6) has a second order accuracy with the following error term

$$\frac{h^2}{12}\left(\frac{\partial^4 u}{\partial x_1^4} + \frac{\partial^4 u}{\partial x_2^4}\right) + \mathcal{O}(h^3).$$

Since solving elliptic equations is the most challenging part for incompressible flow solvers, this chapter mainly deals with this type of equation. Chapter 4 will also investigate the efficiency of developed solvers in case of discretized Poisson's equation (as an elliptic equation).

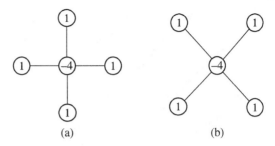

Figure 2.1 Five-point stencil for central difference approximation of Laplace's operator: (a) standard stencil, (b) skewed five-point stencil.

(a)   (b)

2.4 Sparse Matrices

As pointed in the introduction to the chapter, discretization of partial differential equations on structured and unstructured grids leads to production of a system of equations with a large and sparse matrix. Sparse matrix refers to a matrix with few non-zero elements for which special methods are required to store them (Montagne and Ekambaram, 2004). Sparse matrices are typically divided into two major groups: structured and unstructured. In structured sparse matrices, non-zero elements have special patterns and are located on some limited diagonals while no specific pattern of non-zero elements can be found in unstructured sparse matrices. Some examples of sparse matrices along with their storing techniques are discussed later in this chapter. These benchmark matrices will be used in Chapter 4 for comparing the Krylov subspace solvers. A special section of Chapter 4 will be also devoted to comparison of the efficiency of the storage schemes.

2.4.1 Benchmark Problems for Comparing Solvers

To compare the developed solvers, their capabilities must be tested in a wide range of different problems. There are two main approaches here. The first method is to compile sparse matrices in a standard format. Matrices that are introduced here are selected from the Harwell-Boeing sparse matrix collection (Duff, 1992) and the Matrix Market (Boisvert et al., 1997), and in Chapter 4, approaches in which they are used to evaluate the capabilities of solvers are expressed. The second method is the development of subroutines that can be used to extract sparse matrices. In Chapter 4 a sample problem is considered that uses these subroutines. Here some patterns of benchmark sparse matrices are shown in Figures 2.2 to 2.4.

The sparsity patterns of Figures 2.2 to 2.4 are generated with a subroutine that has this feature to generate desired patterns by receiving matrix data in a special format. Different storage formats of sparse matrices are detailed in the next section. Properties of the introduced sparse matrices are presented in Table 2.1. These features help to judge the use of solvers for similar problems and determine the application scope of each method. It is necessary to introduce some definitions

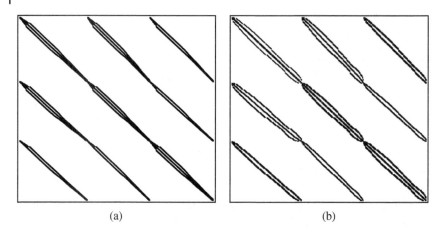

(a)	(b)

Figure 2.2 Sparsity pattern of matrices: (a) PLAT1919, (b) PLAT362.

for some columns of this table. Quantity κ is the condition number of the linear system $Ax = b$ relative to norm ($\|.\|$) (Hoffman, 2001)

$$\kappa(A) = \|A\|\|A^{-1}\|. \tag{2.7}$$

If other standard norms ($\|.\|_p, p = 1, \ldots, \infty$) are used, condition number is indicated by the index p

$$\kappa_p(A) = \|A\|_p \|A^{-1}\|_p.$$

In the case of large matrices, determinant is no longer an appropriate index for sensitivity analysis of linear systems or determining the degree of proximity to a singular status. This is due to the fact that $\det(A)$ is a product of eigenvalues and depends strongly on the scale of the matrix while condition number is independent of matrix scale. For example, determinant of matrix $A = \alpha I$ is equal to α^n and will be very small for $|\alpha| < 1$; this is while $\kappa(A) = 1$ for all standard norms and is independent of matrix scale.

In Table 2.1 we find the diagonally dominant concept that is referred to herein. Matrix A is weakly diagonally dominant if (Meyer, 2000)

$$|a_{jj}| \geq \sum_{\substack{i=1 \\ i \neq j}}^{i=n} |a_{ij}|, j = 1, \ldots, n,$$

and is strictly diagonally dominant if

$$|a_{jj}| > \sum_{\substack{i=1 \\ i \neq j}}^{i=n} |a_{ij}|, j = 1, \ldots, n.$$

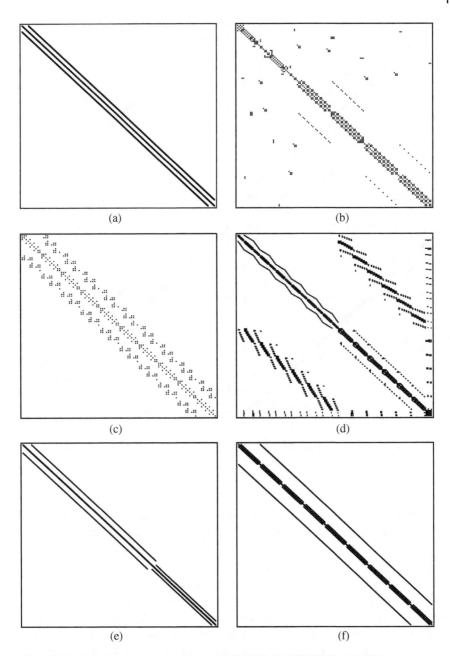

Figure 2.3 Sparsity pattern of matrices: (a) GR3030, (b) BCSSTK22, (c) NOS4,
(d) HOR131, (e) NOS6, (f) NOS7.

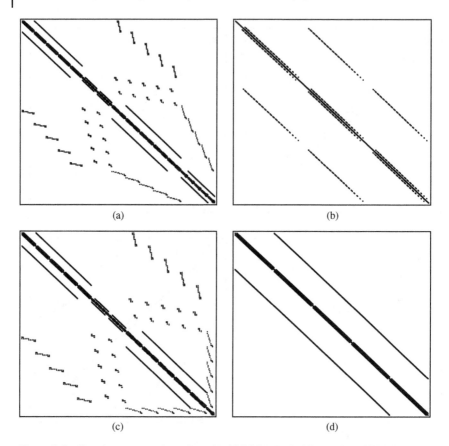

(a)

(b)

(c)

(d)

Figure 2.4 Sparsity pattern of matrices: (a) ORSIRR1, (b) SHERMAN4, (c) ORSIRR2, (d) ORSREG1.

In the fourth column of Table 2.1, reference is made to the type of matrices. Here, it is necessary to clarify the definiteness concept. Consider a symmetric $n \times n$ real matrix A and a non-zero column vector x of n real numbers. In linear algebra, matrix A is said to be positive-definite if the scalar $x^T A x$ is strictly positive. Here x^T denotes the transpose of x. When interpreting Ax as the output of an operator, A, that is acting on an input, x, the property of positive definiteness implies that the output always has a positive inner product with the input, as often observed in physical processes. Matrix A is said to be positive semi-definite or non-negative definite if $x^T A x \geq 0$. Analogous definitions exist for negative definite and negative semi-definite matrices. Matrix A is said to be negative definite if $x^T A x < 0$ and logically negative semi-definiteness or non-positive definiteness is sustained if

Table 2.1 Properties of benchmark matrices.

Matrix	Rank	Non-zero elements	Type	Diagonally dominant	Condition number
PLAT1919	1919	32399	symmetric indefinite	no	1.0e+02
PLAT362	362	5786	symmetric indefinite	no	7.1e+11
BCSSTK22	138	696	symmetric indefinite	no	1.7e+05
GR3030	900	7744	symmetric positive-definite	weakly	3.8e+02
NOS4	100	594	symmetric positive-definite	no	2.7e+03
NOS6	675	3255	symmetric positive-definite	yes	8.0e+06
NOS7	729	4617	symmetric positive-definite	no	4.1e+09
HOR131	434	4710	asymmetric	no	1.3e+05
ORSIRR1	1030	6858	asymmetric	yes	1.0e+02
SHERMAN4	1104	3786	asymmetric	yes	7.2e+03
ORSIRR2	886	5970	asymmetric	yes	1.7e+05
ORSREG1	2205	14133	asymmetric	yes	1.0e+02

$x^T A x \leq 0$. It should be noted that an $n \times n$ symmetric real matrix which is neither positive semi-definite nor negative semi-definite is called indefinite.

Definiteness of a matrix can be also determined based on the eigenvalue sign in its spectrum. Let A be an $n \times n$ symmetric matrix as a special type of Hermitian matrices. Matrix A is positive-definite and semi-definite if and only if all of its eigenvalues are positive and non-negative, respectively; and it is said negative definite and semi-definite if and only if all of its eigenvalues are negative and non-positive, respectively. If matrix A has both positive and negative eigenvalues, it is called indefinite.

2.4.2 Storage Formats of Sparse Matrices

It is not efficient to store all elements of a matrix when most of the elements are zero, so some special storage formats should be introduced for sparse matrices. Here the main purpose is to store non-zero elements such that normal arithmetic operations would still be possible.

2.4.2.1 Coordinate Format

The simplest format of storing sparse matrices is the coordinate format. Data structure in this format is comprised of three vectors: (1) a vector including all real or complex non-zero entries in A (with any order); (2) an integer vector including row indices of non-zero entries; and (3) an integer vector including column indices of non-zero entries. These vectors have length of nz (number of non-zero elements). Suppose that matrix A is

$$A = \begin{pmatrix} a_{11} & a_{12} & 0 & 0 & 0 & 0 \\ 0 & a_{22} & a_{23} & 0 & 0 & 0 \\ 0 & 0 & a_{33} & a_{34} & 0 & 0 \\ a_{41} & 0 & 0 & a_{44} & a_{45} & 0 \\ 0 & a_{52} & 0 & 0 & a_{55} & a_{56} \\ 0 & 0 & a_{63} & 0 & 0 & a_{66} \end{pmatrix}.$$

This matrix in coordinate format is stored as

$$\textbf{AE} \quad \boxed{a_{11} \; a_{12} \; a_{23} \; a_{33} \; a_{34} \; a_{22} \; a_{44} \; a_{55} \; a_{45} \; a_{41} \; a_{63} \; a_{52} \; a_{56} \; a_{66}}$$

$$\textbf{RI} \quad \boxed{1 \; 1 \; 2 \; 3 \; 3 \; 2 \; 4 \; 5 \; 4 \; 4 \; 6 \; 5 \; 5 \; 6}$$

$$\textbf{CI} \quad \boxed{1 \; 2 \; 3 \; 3 \; 4 \; 2 \; 4 \; 5 \; 5 \; 1 \; 3 \; 2 \; 6 \; 6}$$

Here non-zero elements are stored at $\textbf{AE}$, indices of rows and columns of these non-zero elements are stored at $\textbf{RI}$ and $\textbf{CI}$, respectively. In this example, non-zero elements are stored with an arbitrary order. If they are stored by order of their rows, $\textbf{RI}$ vector will contain repeated indices. This feature prompts us to define a more optimal method for storing sparse matrices.

2.4.2.2 Compressed Sparse Row Format

As mentioned in Section 2.4.2.1, storing the indices in $\textbf{RI}$, in row order, leads to the appearance of repetitive indices. It is also possible to replace this vector with another vector in which each element is pointing at the non-zero beginning of each row. New structure will include three vectors: (1) vector $\textbf{AE}$ with length of nz, including non-zero elements of matrix A which are stored row by row; (2) integer vector $\textbf{CI}$ with length nz, including column indices of elements in $\textbf{AE}$; (3) integer vector $\textbf{RP}$ consists of pointers to starting elements of each row in vectors $\textbf{AE}$ and $\textbf{CI}$. Therefore $\textbf{RP}(i)$ is a location of elements in $\textbf{AE}$ and $\textbf{CI}$ which start with the ith row. The length of $\textbf{RP}$ is $n + 1$, where n is the rank of the matrix, and $\textbf{RP}(n + 1)$ is equal to $\textbf{RP}(1) + nz$ which is pointing to the $(n + 1)$th virtual row. So matrix A can also be stored as

$$\textbf{AE} \quad \boxed{a_{11} \; a_{12} \; a_{22} \; a_{23} \; a_{33} \; a_{34} \; a_{41} \; a_{44} \; a_{45} \; a_{52} \; a_{55} \; a_{56} \; a_{63} \; a_{66}}$$

$$\textbf{CI} \quad \boxed{1 \; 2 \; 2 \; 3 \; 3 \; 4 \; 1 \; 4 \; 5 \; 2 \; 5 \; 6 \; 3 \; 6}$$

$$\textbf{RP} \quad \boxed{1 \; 3 \; 5 \; 7 \; 10 \; 13 \; 15}$$

This format is the most common format for storing general sparse matrices which is called the compressed sparse row (CSR) format. Although the coordinate format is more simple and flexible, CSR format is preferred since it requires less memory and is suitable for ordinary applications. This format is used in most toolboxes of sparse matrices as an input format. There are other formats similar to CSR; for instance if columns are used instead of rows, compressed sparse column (CSC) format will be derived.

2.4.2.3 Block Compressed Row Storage Format

With all the advantages of CSR format, it would be more efficient to consider special patterns in sparse matrices in storage format, and one way to do this is to assume a sparse matrix as a combination of $br \times bc$ sub-matrices known as blocks. By this assumption, a limited number of blocks within the main matrix are now dense and non-empty blocks are now considered as new elements in the storing process using CSR format. This format is called block compressed row storage (BCRS). Let $brows = rows/br$ and nzb be the number of blocks with at least one non-zero element. BCRS is shown by three arrays BE, BCI and BRP. All elements in the non-empty blocks are then stored in vector BE of length $nzb \times br \times bc$. The first $br \times bc$ elements are of the first non-zero block, and the next $br \times bc$ elements are of the second non-zero block, etc. The integer array BCI of length nzb stores the block column indices of the non-zero blocks. The integer array BRP of length $(brows + 1)$ stores pointers to the beginning of each block row in the array BCI. Based on what was explained about the CSR format, one can acknowledge that $BRP(brows + 1) = BRP(1) + nzb$. Consider matrix B as follows

$$B = \begin{pmatrix} b_{11} & 0 & b_{13} & 0 & b_{15} & b_{16} & 0 & 0 \\ b_{21} & 0 & 0 & 0 & 0 & 0 & 0 & 0 \\ b_{31} & b_{32} & b_{33} & 0 & b_{35} & 0 & 0 & 0 \\ 0 & 0 & b_{43} & 0 & b_{45} & 0 & 0 & 0 \\ 0 & 0 & 0 & 0 & b_{55} & 0 & 0 & 0 \\ 0 & 0 & 0 & 0 & b_{65} & 0 & 0 & 0 \\ 0 & 0 & 0 & 0 & b_{75} & b_{76} & b_{77} & b_{78} \\ 0 & 0 & 0 & 0 & b_{85} & 0 & b_{87} & 0 \end{pmatrix}.$$

Application of BCRS scheme results in the following three vectors

$$
\textbf{BE} \quad
\begin{array}{l}
b_{11}\ 0\ b_{21}\ 0\ b_{13}\ 0\ 0\ 0\ b_{15}\ b_{16}\ 0\ 0\ b_{31}\ b_{32}\ 0\ 0\ b_{33}\ 0 \\
b_{43}\ 0\ b_{35}\ 0\ b_{45}\ 0\ b_{55}\ 0\ b_{65}\ 0\ b_{75}\ b_{76}\ b_{85}\ 0\ b_{77}\ b_{78}\ b_{87}\ 0
\end{array}
$$

$$\textbf{BCI} \quad 1\ 2\ 3\ 1\ 2\ 3\ 3\ 3\ 4$$

$$\textbf{BRP} \quad 1\ 4\ 7\ 8\ 10$$

As you can observe the matrix entries are stored in vector **BE**, block by block, in the row direction.

2.4.2.4 Sparse Block Compressed Row Storage Format

The sparse block compressed row storage (SBCRS) format is similar in many respects to the BCRS format. The data structure of the format is as follows. The sparse matrix is considered to be constructed from $nb \times nb$ sized blocks. The SBCRS structure includes two parts; a CSR-like structure where the entries are pointers to non-empty nb^2 block arrays representing non-empty $nb \times nb$ sparse blocks in the matrix and the collection of all $nb \times nb$ block arrays. Hence, one can conclude that this is a variation of the BCRS, where the blocks are no longer assumed to be dense and of arbitrary size. All the non-zero elements and their positional data are stored in an array which is named here **BA** which stands for block array. The positional data includes the column and row position of the non-zero entries within the block. Three more arrays are needed to complete the data structure of the SBCRS format. The two arrays **BCI** and **BRP** are similar to BCRS, while the third array **BAP** contains the pointers to the block array **BA**. Once again consider matrix **B** in Section 2.4.2.3. Application of the SBCRS format for storing the matrix **B** results in the following four arrays.

$$
\textbf{BA} \quad
\begin{array}{l}
b_{11}\ b_{13}\ b_{15}\ b_{16}\ b_{21}\ b_{31}\ b_{32}\ b_{33}\ b_{35}\ b_{43}\ b_{45}\ b_{55}\ b_{65}\ b_{75}\ b_{76}\ b_{77}\ b_{78}\ b_{85}\ b_{87} \\
1\ \ 1\ \ 1\ \ 1\ \ 2\ \ 1\ \ 1\ \ 1\ \ 1\ \ 2\ \ 2\ \ 1\ \ 2\ \ 1\ \ 1\ \ 1\ \ 1\ \ 2\ \ 2 \\
1\ \ 1\ \ 1\ \ 2\ \ 1\ \ 1\ \ 2\ \ 1\ \ 1\ \ 1\ \ 1\ \ 1\ \ 1\ \ 1\ \ 2\ \ 1\ \ 2\ \ 1\ \ 1
\end{array}
$$

$$\textbf{BAP} \quad 1\ 2\ 3\ 6\ 8\ 9\ 12\ 14\ 16$$

$$\textbf{BCI} \quad 1\ 2\ 3\ 1\ 2\ 3\ 3\ 3\ 4$$

$$\textbf{BRP} \quad 1\ 4\ 7\ 8\ 10$$

Here, the contents of the foregoing arrays are clarified. It should be noted that the second and third rows of the array **BA** point to the block row and column indices of the corresponding element of the first row. The elements of vector **BAP** are pointers to those elements of the first row of array BA which is the first non-zero entry of the corresponding block when it is read row by row. Strictly speaking, the kth entry of the vector BAP is filled with the column index of entry b_{ij} which is addressed as the first non-zero element of the kth block when it is read row by row. The vectors BCI and BRP, as mentioned above, have the same definition as in BCRS format. Visual description of this format is given in Figure 2.5 (Smailbegovic et al., 2005; Shahnaz and Usman, 2011).

2.4.2.5 Modified Sparse Row Format

Since the diagonal elements of many matrices are non-zero and more accessible than other non-zero elements of a matrix, it is possible to store them separately. Modified sparse row (MSR) format benefits from this idea and needs only two vectors; a real vector AA and an integer vector JA. The first n positions of vector AA are devoted to diagonal elements of the matrix. The position $n + 1$ remains unused, but in some cases it is used to transmit some information of the matrix. From position $n + 2$ onwards, non-zero off-diagonal elements are stored and for any element $AA(k)$, integer number $JA(k)$ denotes the corresponding column index. The first $n + 1$ elements of vector JA belong to pointers to the starting off-diagonal elements of each row in AA and JA (Saad, 2011). Once again consider the sparse matrix A of Section 2.4.2.1. Applying the MSR format results in the following two vectors

$$AA \quad \boxed{a_{11} \; a_{22} \; a_{33} \; a_{44} \; a_{55} \; a_{66} \; * \; a_{12} \; a_{23} \; a_{34} \; a_{41} \; a_{45} \; a_{52} \; a_{56} \; a_{63}}$$

$$JA \quad \boxed{8 \; 9 \; 10 \; 11 \; 13 \; 15 \; 15 \; 2 \; 3 \; 4 \; 1 \; 5 \; 2 \; 6 \; 3}$$

Here, the star ($*$) represents the position that is either left unused or used to transmit some matrix information and $JA(n + 1)$ also points to the start of a virtual row. Strictly speaking, $JA(n + 1)$ is set to be equal to $nz + dz + 1$; where nz and dz are the number of non-zero and diagonal zero elements, respectively.

2.4.2.6 Diagonal Storage Format

Sparse matrices with non-zero elements on a limited number of diagonals are referred to as band matrices. Diagonal elements can be stored in a two dimensional array $DIA(1 : n, 1 : nd)$ where nd is the number of diagonals. Deviation of the diagonals from the main diagonal is also stored in the array $OFFSET(1 : nd)$. So element $a_{i,i+OFFSET(j)}$ from the main matrix is located in (i,j) of array DIA, hence

$$DIA(i,j) \leftarrow a_{i,i+OFFSET(j)}.$$

This format is called diagonal. Diagonals of the matrix can be stored in any order in columns of array DIA but since the main diagonal deals with more arithmetic

Red color denotes nonzeros

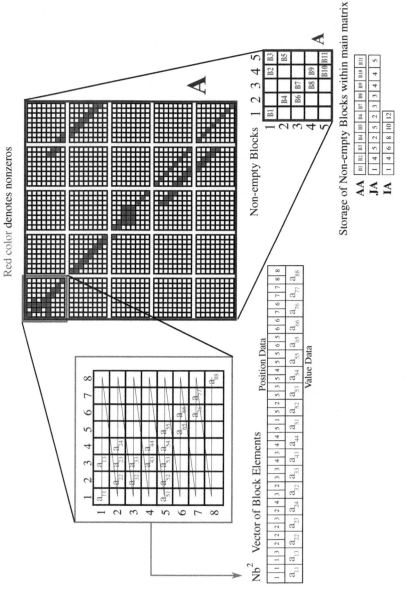

Figure 2.5 Visual description of Block Compressed Row Storage format.

operations, it is preferred that it is stored in the first column of **DIA**. It should be noted that the number of non-zero elements in all diagonals except the main diagonal is less than n, thus, some elements of **DIA** are left unused. It can be understood that diagonal format for storing the band matrices is highly efficient and unused locations in **DIA** are minimized (Pissanetzky, 1984). It is obvious that the implementation of this format in the case of sparse matrices with no specific pattern of non-zero elements will degrade efficiency of matrix arithmetic operations. Storing the matrix **A** from Section 2.4.2.1 using this format will result in these arrays

$$
\mathbf{DIA}
\begin{vmatrix}
a_{11} & a_{12} & * \\
a_{22} & a_{23} & * \\
a_{33} & a_{34} & * \\
a_{44} & a_{45} & a_{41} \\
a_{55} & a_{56} & a_{52} \\
a_{66} & * & a_{63}
\end{vmatrix}
\qquad
\mathbf{OFFSET}\ \boxed{0\ \ 1\ \ -3}
$$

2.4.2.7 Compressed Diagonal Storage Format

In some cases, finite element or finite difference discretization on a tensor product grid gives rise to sparse matrices with non-zero diagonal elements (Sanjuan et al., 2016) that provide the possibility of using compressed diagonal storage (CDS) format, in which sub-diagonal and super-diagonal elements of the matrix are stored in consecutive locations. As a result, there will be no need for extra arrays to store row and column locations of elements. A matrix is called banded when two non-negative integers of p and q can be defined as left and right half-bandwidth, respectively, such that (Bai et al., 2000)

$$c_{ij} \neq 0 \Leftrightarrow i - p \leq j \leq i + q$$

and to store these bands in array **VAL**$(1 : n, -p : q)$ one can consider the following assignment

$$\mathbf{VAL}(i, j) \leftarrow c_{i, i+j}.$$

Here is an example of a band matrix **C** and the corresponding array **VAL**

$$
C = \begin{pmatrix}
c_{11} & c_{12} & 0 & 0 \\
c_{21} & c_{22} & c_{23} & 0 \\
0 & c_{32} & c_{33} & c_{34} \\
0 & 0 & c_{43} & c_{44}
\end{pmatrix}
\qquad
\mathbf{VAL}
$$

$-p = -1$	0	$q = 1$
0	c_{11}	c_{12}
c_{21}	c_{22}	c_{23}
c_{32}	c_{33}	c_{34}
c_{43}	c_{44}	0

As can be seen, the extra zeros inserted into **VAL** are in fact imaginary elements of their corresponding bands which extend the number of band elements to matrix rank ($n = 4$).

2.4.2.8 Ellpack-Itpack Format

Another format that is commonly used in vector machines is the Ellpack-Itpack format. The base assumption in this format is that the maximum number of non-zero elements in each row is nd where this number is small. Now two two-dimensional arrays with size $n \times nd$ are used (one of them is real and the other is integer). Non-zero elements of each row are stored in the corresponding row of array **ELEMENT**. Since non-zero elements in some rows are less than nd, consequently blank locations are filled with zeros. Integer array **COL** stores the column number of each corresponding element in array **ELEMENT**. For the column position of zeros in array **ELEMENT**, a non-repeated number between 1 to n is chosen. The matrix **A** from Section 2.4.2.1 in this format is represented as follows

$$
\textbf{\textit{ELEMENT}}
\begin{bmatrix}
a_{11} & a_{12} & 0 \\
a_{22} & a_{23} & 0 \\
a_{33} & a_{34} & 0 \\
a_{41} & a_{44} & a_{45} \\
a_{52} & a_{55} & a_{56} \\
a_{63} & a_{66} & 0
\end{bmatrix}
\qquad
\textbf{\textit{COL}}
\begin{bmatrix}
1 & 2 & 1 \\
2 & 3 & 2 \\
3 & 4 & 3 \\
1 & 4 & 5 \\
2 & 5 & 6 \\
3 & 6 & 4
\end{bmatrix}
$$

This format is not recommended for storing matrices with non-uniform distribution of non-zero elements on rows (Kincaid and Young, 1984). In Chapter 4, the effects of some storage formats on convergence CPU-time are studied.

2.4.3 Redefined Matrix–Vector Multiplication

Since compressed formats will not give explicit representation of matrices, it is necessary to redefine corresponding operations to use these formats. Matrix–vector multiplication is one of the most important operations which is redefined in Krylov subspace methods. Algorithms performing this redefined operation for CSR and diagonal formats are represented as Algorithm 2.1 and 2.2 (Chen et al., 2018).

Algorithm 2.1 Redefined matrix–vector multiplication in CSR format

---

1: **Input:** AE, CI, RP and x (in $Ax = b$)
2: **Output:** b
3: **for** $i = 1, \ldots, n$ **do**
4:     $sum = 0$;
5:     **for** $j = 1, \ldots, RP(i) - 1$ **do**
6:         $sum = AE(j)x(CI(j)) + sum$;
7:     **end for**
8:     $b(i) = sum$;
9: **end for**

Algorithm 2.2 Redefined matrix–vector multiplication in diagonal format

---

1: **Input:** n, nd, DIA, $OFFSET$ and x (in $Ax = b$)
2: **Output:** b
3: **for** $i = 1, \ldots, nd$ **do**
4:     $d = OFFSET(i)$;
5:     $r_0 = max(1, 1 - d)$;
6:     $r_1 = min(n, n - d)$;
7:     $c_0 = max(1, 1 + d)$;
8:     **for** $r = r_0, \ldots, r_1$ **do**
9:         $c = r - r_0 + c_0$;
10:         $x(r) = x(r) + DIA(r, i)x(c)$;
11:     **end for**
12: **end for**

Exercises

2.1 Consider Poisson's equation $u_{xx} + u_{yy} = f(x, y)$

    (a) Discretize the equation using the second order central difference approximation and store the resulted five-point stencil matrix in the coordinate format. Develop a FORTRAN or C programs to do this.

    (b) Apply the subroutine of Appendix A.4 for showing the matrix pattern.

    (c) Apply the subroutines of Appendices A.1.1 to A.1.4 and execute the process of matrix conversion from coordinate format to CSR and CSR to MSR, Ellpack-Itpack and diagonal formats.

2.2 Design separate algorithms for converting the CSR to BCRS, SBCRS and CDS formats and develop their corresponding FORTRAN or C programs.

2.3 Consider the five-point stencil matrix of Exercise 1 and develop a FORTRAN or C program to calculate the required storage memory of all formats of Section 2.4.2 versus matrix dimension and compare them.

2.4 Design separate algorithms for matrix–vector multiplication when the matrix is stored in BCRS and SBCRS formats and develop their corresponding FORTRAN or C programs.

2.5 For sparse matrices where their elements have low value diversity, e.g., five-point stencil matrices,
  (a) Design an algorithm for storing them in a more compressed storage format compared to what you have learned so far.
  (b) Design an algorithm for their matrix–vector multiplication operation.
  (c) Develop a FORTRAN or C programs for your proposed algorithms for the two previous parts.

2.6 Design an algorithm in which the matrix in CSR format is converted to its transposed form in a desired compressed format. Develop a FORTRAN or C program to do this.

2.7 Repeat Exercises 1 and 3 when a fourth order central difference approximation is applied and we have a nine-point stencil matrix.

2.8 Consider the Poisson's equation $u_{xx} + u_{yy} + u_{zz} = f(x,y,z)$ and discretize the equation using the second order central difference approximation and store the resulting seven-point stencil matrix in the coordinate format. Develop a FORTRAN or C program to do this. Repeat parts (b) and (c) of Exercise 1.

2.9 Consider the seven-point stencil matrix of Exercise 8 and respond to the demands of Exercise 3.

3

Theory of Krylov Subspace Methods

3.1 Introduction

The theoretical foundations of the most important available methods known for solving large and spars linear systems are disscused in this chapter. These methods are based on orthogonal or oblique projection processes. These subspaces are formed by vectors in the form $p(A)v$, where p is a polynomial. In fact, in these methods, we try to estimate $A^{-1}b$ with $p(A)b$ (Saad, 1996; Meyer, 2000). This chapter focuses solely on the mathematical foundations, theory and derivation of Krylov subspace methods, and finally an algorithm of each method is independently presented.

3.2 Projection Methods

Most of the common iteration based methods utilize projection process for solving the large linear system of equations. A projection process represents a way of extracting an approximation to the solution of a linear system from a subspace. This section explains these techniques in a general framework and presents some theories.

At first, it seems necessary to provide a definition of subspace. A subspace of $\mathbb{C}^n$ is a subset of $\mathbb{C}^n$ that is also a complex vector space. The set of all linear combinations of a set of vectors $G = \{a_1, a_2, ..., a_q\}$ of $\mathbb{C}^n$ is a vector subspace called the linear span of G,

$$\text{span}\{G\} = \text{span}\{a_1, a_2, ..., a_q\} = \{z \in \mathbb{C}^n | z = \sum_{i=1}^{q} \alpha_i a_i; \{\alpha_i\}_{i=1,...,q} \in \mathbb{C}^q\}.$$

Krylov Subspace Methods with Application in Incompressible Fluid Flow Solvers, First Edition.
Iman Farahbakhsh.
Companion Website: www.wiley.com/go/Farahbakhs/KrylovSubspaceMethods

If the vectors a_i are linearly independent, then each vector of span$\{G\}$ can be considered as a unique expression as a linear combination of the vectors a_i. The set G is then called a basis of the subspace span$\{G\}$ (Saad, 1996).

Consider the following linear system

$$Ax = b \tag{3.1}$$

where A is a real $n \times n$ matrix. The essential idea of the projection method is extraction of an approximate solution from a subspace in $\mathbb{R}^n$ for the foregoing system of equations. If $\mathcal{K}$ is a subspace of dimension m, it is clear that m constraints are required for obtaining an approximate solution from this subspace. In other words, for describing these constraints, m independent orthogonality conditions must be imposed. Strictly speaking, the residual vector $b - Ax$ is constrained to be orthogonal to m linearly independent vectors. Here, another m dimensional subspace $\mathcal{L}$ must be defined which will be called a subspace of constraints. This simple framework is common to many different mathematical methods and is known as the Petrov–Galerkin conditions (Van der Vorst, 2003).

There are two broad classes of projection methods: orthogonal and oblique. In an orthogonal projection technique, the subspace $\mathcal{L}$ is the same as $\mathcal{K}$. In an oblique projection method, $\mathcal{L}$ is different from $\mathcal{K}$ and may be totally unrelated to it. This distinction is rather significant and results in different types of algorithms (Saad, 1996).

Let A be a $n \times n$ real matrix and $\mathcal{K}$ and $\mathcal{L}$ be two m dimensional subspaces in $\mathbb{R}^n$. A projection technique is a process which finds an approximate solution $\tilde{x}$ to Equation (3.1) by applying the conditions that $\tilde{x}$ belongs to $\mathcal{K}$ and that the new residual vector be orthogonal to $\mathcal{L}$. Let the initial guess $x^{(0)}$ for the solution vector be available, then the approximation must be sought in the affine space $x^{(0)} + \mathcal{K}$ instead of the homogeneous vector space $\mathcal{K}$. The mentioned conditions can be written as follows

$$\text{Find } \tilde{x} \in \mathcal{K}, \text{ such that } b - A\tilde{x} \perp \mathcal{L}, \tag{3.2}$$

$$\text{Find } \tilde{x} \in x^{(0)} + \mathcal{K}, \text{ such that } b - A\tilde{x} \perp \mathcal{L}. \tag{3.3}$$

If $\tilde{x}$ is written as $\tilde{x} = x^{(0)} + \delta$ and initial residual vector is defined as $r^{(0)} = b - Ax^{(0)}$, the orthogonality relation can be expressed as

$$b - A(x^{(0)} + \delta) \perp \mathcal{L} \text{ or } r^{(0)} - A\delta \perp \mathcal{L}.$$

In other words, the approximate solution can be defined as

$$\tilde{x} = x^{(0)} + \delta, \ \delta \in \mathcal{K} \tag{3.4}$$

$$(r^{(0)} - A\delta, w) = 0, \ \forall w \in \mathcal{L}. \tag{3.5}$$

The orthogonality condition of Equation (3.5) must be imposed on the new residual vector $r_{\text{new}} = r^{(0)} - A\delta$ as illustrated in Figure 3.1. Most standard methods

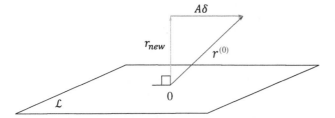

Figure 3.1 Geometrical interpretation of orthogonalization condition.

use this projection process for finding the solution, successively. Typically, in the next step of the projection method, new subspaces $\mathcal{K}$ and $\mathcal{L}$ are used and the last approximate solution is referred to as an initial guess $x^{(0)}$. Projection techniques constitute a monolithic framework for most of the well-known methods of scientific computing. Actually, all of the stationary iterative methods can be typically considered as projection methods. Generally, in the projection process, an approximate solution can be expressed with m degrees of freedom (subspace $\mathcal{K}$) and n constraints (subspace $\mathcal{L}$).

Suppose that $V = [v_1, v_2, ..., v_m]$ and $W = [w_1, w_2, ..., w_m]$ are $n \times m$ matrices whose column vectors form the basis of $\mathcal{K}$ and $\mathcal{L}$, respectively. If the approximate solution is written as

$$x = x^{(0)} + Vy,$$

with imposing the orthogonality condition one can find that

$$(b - A(x^{(0)} + Vy), W) = (r^{(0)} - AVy, W) = 0,$$

which directly results in

$$W^T AVy = W^T r^{(0)}.$$

If $W^T AV$ is assumed to be a non-singular $m \times m$ matrix, y can be expressed as

$$y = (W^T AV)^{-1} W^T r^{(0)}$$

and consequently the following expression for the approximate solution $\tilde{x}$ results in

$$\tilde{x} = x^{(0)} + V(W^T AV)^{-1} W^T r^{(0)}. \tag{3.6}$$

On the basis of the foregoing explanations, a primary algorithm of the projection method, Algorithm 3.1, will be presented here. It should be noted that the non-singularity of the matrix $W^T AV$ is a necessary condition for the existence of an approximate solution, which is not guaranteed to be true even when A is non-singular.

Algorithm 3.1 Primary Projection Method

1: **Initialization:**
2:    $x^{(i)} = x^{(0)}$
3: **while** The convergence criteria is not met **do**
4:    Select a pair of subspaces $\mathcal{K}$ and $\mathcal{L}$
5:    Choose bases $V = [v_1, v_2, \ldots, v_m]$ and $W = [w_1, w_2, \ldots, w_m]$
      for $\mathcal{K}$ and $\mathcal{L}$
6:    $r^{(i)} = b - Ax^{(i)}$
7:    $y^{(i)} = (W^T A V)^{-1} W^T r^{(i)}$
8:    $x^{(i+1)} = x^{(i)} + V y^{(i)}$
9: **end while**

3.3 Krylov Subspace

As mentioned in the previous section, the projection method for the solution of a linear system, is a process in which an approximate solution $x^{(m)}$ is sought from an m dimensional subspace $x^{(0)} + \mathcal{K}_m$ by imposing the Petrov–Galerkin orthogonality condition

$$b - Ax^{(m)} \perp \mathcal{L}_m,$$

where $\mathcal{L}_m$ is another m dimensional subspace. It should be noted that $x^{(0)}$ is an arbitrary initial guess to the solution. A Krylov subspace method is a method that its search subspace $\mathcal{K}_m$ is a Krylov subspace

$$\mathcal{K}_m(A, r^{(0)}) = \text{span}\{r^{(0)}, Ar^{(0)}, A^2 r^{(0)}, \ldots, A^{m-1} r^{(0)}\},$$

where $r^{(0)} = b - Ax^{(0)}$. For simplicity, when there is no ambiguity, one can show $\mathcal{K}_m(A, r^{(0)})$ by $\mathcal{K}_m$. The various choices of the subspace $\mathcal{L}_m$ and the methods in which the system of equations is preconditioned results in different versions of Krylov subspace methods. Preconditioning the systems will be addressed as an independent topic in Chapter 4. With respect to former statements in the topic of finding an approximate solution from a subspace, one can easily see that a derived approximate solution from the Krylov subspace is expressed as

$$A^{-1}b \approx x^{(m)} = x^{(0)} + q_{m-1}(A)r^{(0)},$$

where q_{m-1} is a certain polynomial of degree $m - 1$. Consider a situation where $x^{(0)} = 0$, then

$$A^{-1}b \approx q_{m-1}(A)b.$$

In spite of the fact that in all methods the same type of polynomial approximations are provided, the choice of $\mathcal{L}_m$, i.e., the constraints used to build these approximations, will have a significant effect on the iterative technique. There are two extensive choices for $\mathcal{L}_m$ which give rise to the well-known techniques. The first is simply expressed as $\mathcal{L}_m = \mathcal{K}_m$ and the minimum-residual variation that is $\mathcal{L}_m = A\mathcal{K}_m$. The second class of methods is based on defining $\mathcal{L}_m$ to be a Krylov subspace method associated with A^T, that is, $\mathcal{L}_m = \mathcal{K}_m(A^T, r^{(0)})$.

3.4 Conjugate Gradient Method

This section is devoted to briefly introducing the conjugate gradient method (CG) which was initially developed by Hestenes and Stiefel (1952). In the absence of round-off error, this method can obtain an exact solution for a system of equations with $n \times n$ non-singular symmetric matrix A in at most N iterations (McDonough, 2008). It should be noted that when the method was first introduced it was well-known as a direct solver. The computational cost of this method is of order N^2, $\mathcal{O}(N^2)$, which is competitive with common SOR which was essentially the only viable alternative at that time. Unfortunately, round-off errors dramatically reduce the performance and mean that one cannot use this method immediately. Initially, basic theories are presented in conjunction with the minimization with steepest descent method. Further, these theories are used to derive the conjugate gradient method.

3.4.1 Steepest Descent Method

At first, it is assumed that the matrix A is a $n \times n$ symmetric positive-definite matrix. One can begin derivation of the steepest descent method by constructing the following quadratic form (Meyer, 2000)

$$F(x) = \frac{1}{2}(x, Ax) - (b, x), \tag{3.7}$$

which is the equivalent to the system $Ax = b$. Inasmuch as A is positive-definite, it is deduced that $F(x)$ has a minimum. In spite of the fact that A is positive-definite, it can be seen that $Ax = b$ corresponds to a critical point of $F(x)$; namely, minimizing (or more generally, finding the critical point of) $F(x)$ is equivalent to solving $Ax = b$, since there is a minimum (or critical point) value where $\nabla_x F(x) = 0$ and

$$\nabla_x F(x) = Ax - b.$$

To minimize F, starting with the approximate value of $x^{(i)}$, the new approximate value can be obtained by moving in the direction of the negative gradient. This process is shown in the following formula

$$x^{(i+1)} = x^{(i)} - \lambda_i \nabla F(x^{(i)}) = x^{(i)} + \lambda_i r^{(i)}. \tag{3.8}$$

As already mentioned

$$r^{(i)} = b - Ax^{(i)},\tag{3.9}$$

and λ_i is a length step given by

$$\lambda_i = \frac{(r^{(i)}, r^{(i)})}{(r^{(i)}, Ar^{(i)})}.\tag{3.10}$$

This relation can be obtained by substituting Equation (3.8) into Equation (3.7), using Equation (3.9) to eliminate $x^{(i)}$, and then minimizing the result with respect to the scalar λ_i (Meurant, 2006). Figure 3.2 shows a schematic of the convergence process of the steepest descent method for a 2×2 system of equations. It should be noted that eccentricity of elliptical contours in this figure is very low and this represents that matrix A is well-conditioned. In this case, the largest and smallest eigenvalues of this matrix are close to each other, and the matrix condition number $\kappa(A)$ should be close to unity. The ratio of the smallest and largest eigenvalues can be used as a scale for the ellipse axes ratio.

It is observed that this method is well answered in this case, and this is due to the fact that the gradient direction in the first step adopts a path that is very close to the neighborhood of the minimum point. In addition, because the eccentricity of the ellipsis is not too large, the directions of the gradient are proper search directions. When the eccentricity of the ellipsis is high and the condition number is large, this method does not work well; see Figure 3.3. In this figure the steepest descent convergence trajectory and the trajectory corresponding to the conjugate gradient

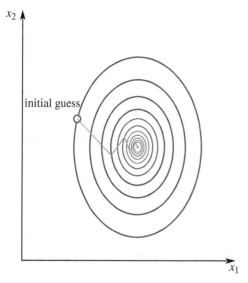

Figure 3.2 Convergence trace of steepest descent method for an equation system with a 2×2 coefficient matrix. The initial guess is illustrated with a hollow circle and the center of the ellipses shows the approximate solution with minimum residual.

Figure 3.3 Convergence trace of steepest descent method (solid line) and conjugate gradient method (dashed line) for an equation system with a 2 × 2 coefficient matrix. The initial guess is illustrated with a hollow circle and center of ellipses shows the approximate solution with minimum residual.

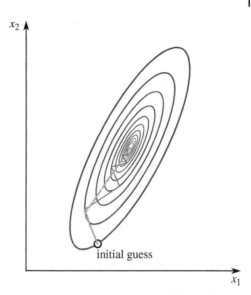

method are shown. It is clear in this figure that the conjugate gradient method has an extraordinary superiority to the steepest descent method. The reason for the superiority of the conjugate gradient method is that it uses information to select the search directions, which is specific to the matrix of the equations system. Hence, this method is not affected by the adverse effects of the eccentricity of the elliptic search areas. The convergence trajectory of the steepest descent method in a region with stretched contours has intense fluctuations. While the conjugate gradient method uses the natural directions which are given by the eigenvectors of the system matrix, the steepest descent method follows the gradient direction, which is the reason for the intense fluctuations (McDonough, 2008). Algorithm 3.2 shows the process in which steepest descent method finds the approximate solution.

Algorithm 3.2 Steepest Descent Method

1: **Initialization:**
2: $\quad x^{(i)} = x^{(0)}$
3: **while** The convergence criteria is not met **do**
4: $\quad r^{(i)} = b - Ax^{(i)}$
5: $\quad \alpha_i = (r^{(i)}, r^{(i)})/(Ar^{(i)}, r^{(i)})$
6: $\quad x^{(i+1)} = x^{(i)} + \alpha_i r^{(i)}$
7: **end while**

3.4.2 Derivation of Conjugate Gradient Method

The conjugate gradient method can be extracted in a variety of ways. Here, a method similar to that used for the steepest descent method is used in more detail. Therefore, as previously mentioned, the basic idea is to minimize the function

$$F(x) = \frac{1}{2}(x, Ax) - (b, x).$$

The matrix A is a symmetric positive-definite matrix so that the minimum value for the above function can be guaranteed. The process of approaching the minimum value (starting from the initial guess $x^{(0)}$) can be expressed by the following formula (Barrett et al., 1994; McDonough, 2008; Meurant, 2006)

$$x^{(i+1)} = x^{(i)} + \alpha_i p^{(i)}. \tag{3.11}$$

Here, $p^{(i)}$ and α_i are search direction and step length, respectively, which are obtained using iteration data from step i. The structure of iterations in the conjugate gradient method and the steepest descent method are very similar, with the difference that the conjugate gradient method is used to search in such a way that the convergence trajectory does not come with severe fluctuations in cases where the matrix condition number is large, see Figure 3.3. The search direction of the conjugate gradient method in each iteration step is expressed as follows

$$p^{(i)} = r^{(i)} + \beta_i p^{(i-1)}, \tag{3.12}$$

where, the quantity β_i is such that

$$(p^{(i)}, Ap^{(i-1)}) = 0. \tag{3.13}$$

By substituting Equation (3.12) into Equation (3.13) we have

$$(r^{(i)}, Ap^{(i-1)}) + \beta_i(p^{(i-1)}, Ap^{(i-1)}) = 0,$$

or in other words

$$\beta_i = -\frac{(r^{(i)}, Ap^{(i-1)})}{(p^{(i-1)}, Ap^{(i-1)})}$$

It is clear that the matrix A must be positive-definite to guarantee the existence of the β_i (McDonough, 2008).

There are two different, but equivalent, methods for determining α_i; one uses the orthogonality principle of successive residual vectors and the other minimizes $F(x)$ with respect to α_i. The first method is obvious, so the second method is considered to simultaneously express missing details of the steepest descent method. Hence, substituting Equation (3.11) into Equation (3.7) results in

$$F(x^{(i+1)}) = \frac{1}{2}((x^{(i)} + \alpha_i p^{(i)}), A(x^{(i)} + \alpha_i p^{(i)})) - (b, (x^{(i)} + \alpha_i p^{(i)}))$$

$$= \frac{1}{2}[(x^{(i)}, Ax^{(i)}) + \alpha_i(p^{(i)}, Ax^{(i)}) + \alpha_i(x^{(i)}, Ap^{(i)}) + \alpha_i^2(p^{(i)}, Ap^{(i)})]$$

$$- (b, x^{(i)}) - \alpha_i(b, p^{(i)}).$$

To minimize this relation with respect to α_i we set

$$\frac{\partial F(x^{(i+1)})}{\partial \alpha_i} = 0,$$

which results in

$$\frac{1}{2}[(p^{(i)}, Ax^{(i)}) + (x^{(i)}, Ap^{(i)}) + 2\alpha_i(p^{(i)}, Ap^{(i)})] = (b, p^{(i)}).$$

Since the matrix A is symmetric, the above relation can be simplified in the following way

$$(p^{(i)}, Ax^{(i)}) + \alpha_i(p^{(i)}, Ap^{(i)}) = (b, p^{(i)}).$$

Since $Ax^{(i)} = b - r^{(i)}$, the above relation leads to

$$\alpha_i = \frac{(p^{(i)}, r^{(i)})}{(p^{(i)}, Ap^{(i)})}.$$

The previous formulas can be obtained in another form. Using Equations (3.9) and (3.11), we can show that the residual vector is updated in each iteration using the following equation

$$r^{(i+1)} = r^{(i)} - \alpha_i Ap^{(i)}. \tag{3.14}$$

Then, using the above relation and the orthogonality principle of the residuals vector, we can write

$$(r^{(i)} - \alpha_i Ap^{(i)}, r^{(i)}) = 0,$$

and we can conclude that

$$\alpha_i = \frac{(r^{(i)}, r^{(i)})}{(Ap^{(i)}, r^{(i)})}.$$

With similar manipulations, a simpler formula can be obtained for β_i, i.e.

$$\beta_i = \frac{(r^{(i+1)}, r^{(i+1)})}{(r^{(i)}, r^{(i)})}.$$

The formulas obtained for calculating α_i and β_i are equivalent to each other (Hageman and Young, 1981), and to obtain them, the identities obtained by Hestenes and Stiefel (1952) have been used. These identities are

$$(r^{(i)}, r^{(j)}) = 0 \ \forall i \neq j$$
$$(p^{(i)}, Ap^{(j)}) = 0 \ \forall i \neq j$$
$$(r^{(i)}, Ap^{(j)}) = 0 \ \forall i \neq j, \ i \neq j + 1.$$

The algorithm of the conjugate gradient method, Algorithm 3.3, is presented as follows

---

Algorithm 3.3 Conjugate Gradient Method

1: **Initialization:**
2:     $r^{(0)} = b - Ax^{(0)}$
3:     $p^{(0)} = r^{(0)}$
4: **while** The convergence criteria is not met **do**
5:     $\alpha_i = (p^{(i)}, r^{(i)})/(p^{(i)}, Ap^{(i)})$
6:     $x^{(i+1)} = x^{(i)} + \alpha_i p^{(i)}$
7:     $r^{(i+1)} = b - Ax^{(i+1)}$
8:     $\beta_{i+1} = -(r^{(i+1)}, Ap^{(i)})/(p^{(i)}, Ap^{(i)})$
9:     $p^{(i+1)} = r^{(i+1)} + \beta_{i+1} p^{(i)}$
10: **end while**

3.4.3 Convergence

In spite of the fact that accurate prediction of the iteration based method convergence is difficult, but it is feasible to obtain its range. In the conjugate gradient method, one can find a range for error using spectral condition number of matrix A. If $\lambda_{\max}$ and $\lambda_{\min}$ are the largest and smallest eigenvalues of a symmetric positive-definite matrix such as A, the spectral condition number of the matrix A is

$$\kappa(A) = \frac{\lambda_{\max}(A)}{\lambda_{\min}(A)}.$$

If A is a symmetric positive-definite matrix and $\hat{x}$ is the exact solution of the linear system $Ax = b$, for the conjugate gradient method, it can be shown that

$$\|x^{(i)} - \hat{x}\|_A \leq 2\zeta^i \|x^{(0)} - \hat{x}\|_A,$$

where $\zeta = (\sqrt{\kappa} - 1)/(\sqrt{\kappa} + 1)$ and $\|y\|_A^2 = (y, Ay)$ (Golub and Van Loan, 1989; Kaniel, 1966). Hence, we find that the number of iterations needed to achieve a relative reduction in error is proportional to $\sqrt{\kappa}$ (Barrett et al., 1994).

In some cases, the practical application of the above method is not complicated to determine the range of convergence. For example, elliptic PDEs generally lead to linear system of equations with a coefficient matrix A in which $\kappa(A) = \mathcal{O}(h^{-2})$ and h is the discretization distance. The order of condition number is independent of the finite difference approximation order and spatial dimensions of the considered problem (Axelsson and Barker, 1984). Therefore, for a conjugate gradient method without preconditioning, it is expected that the number of iterations is proportional to h^{-1}. Other results have also been obtained regarding the behavior of the conjugate gradient method. If the extreme eigenvalues of the matrix A are well-posed, then we can see the superlinear convergence of the conjugate gradient method, that is, the convergence rate increases in each iteration. This phenomenon can be explained in such a way that, in the conjugate gradient method, the error components are eliminated in the direction of the eigenvectors corresponding to the extreme eigenvalues (Barrett et al., 1994); then, this process is repeated in the next steps, and in each step, the previous extreme eigenvalues are not considered, and the convergence rate is dependent on a reduced system with a much smaller condition number (Van der Sluis and van der Vorst, 1986). The effect of preconditioners on reducing the condition number and the arrangement of eigenvalues can be deduced by studying the eigenvalues obtained by the Lanczos process. The conjugate gradient method involves one matrix–vector product, three vector updates, and two inner products per iteration. The conjugate gradient method can be implemented in different forms, but the structure of each one is the same (Reid, 1971).

3.5 Minimal Residual Method

The conjugate gradient method can be considered as a special type of Lanczos method, which is used for solving the systems with symmetric positive-definite matrices. Minimal residual method (MINRES method) is another type of Lanczos method that can be used for solving the systems with symmetric indefinite matrices (Barrett et al., 1994). The vector sequences in the conjugate gradient method correspond to a factorization of a tridiagonal matrix similar to the coefficient matrix. Therefore, a breakdown of the algorithm can occur corresponding to a zero pivot if the matrix is indefinite. Furthermore, for indefinite matrices the minimization property of the conjugate gradient method is no longer well-defined. The MINRES method is a variant of the CG method that avoids the LU-factorization and does not suffer from breakdown. The MINRES method minimizes the Euclidean norm of residual.

When A is not positive-definite, but symmetric, one can construct an orthogonal basis for the Krylov subspace by three term recurrence relations. Eliminating the search directions in Equation (3.12) and Equation (3.14) results in a recurrence

$$Ar^{(i)} = r^{(i+1)}t_{i+1,i} + r^{(i)}t_{i,i} + r^{(i-1)}t_{i-1,i},$$

which in matrix form is

$$AR_i = R_{i+1}\overline{T}_i,$$

where $\overline{T}_i$ is an $(i+1) \times i$ tridiagonal matrix. However, we can minimize the residual in the Euclidean norm by obtaining

$$x^{(i)} \in \{r^{(0)}, Ar^{(0)}, ..., A^{i-1}r^{(0)}\}, \; x^{(i)} = R_i\overline{y},$$

that minimizes

$$\|Ax^{(i)} - b\|_2 = \|AR_i\bar{y} - b\|_2 = \|R_{i+1}\overline{T}_iy - b\|_2.$$

Now we take advantage of the fact that if $D_{i+1} \equiv \mathrm{diag}(\|r^{(0)}\|_2, \|r^{(1)}\|_2, \|r^{(i)}\|_2)$, then $R_{i+1}D_{i+1}^{-1}$ can be considered as an orthonormal transformation with respect to the current Krylov subspace

$$\|Ax^{(i)} - b\|_2 = \|D_{i+1}\overline{T}_iy - \|r^{(0)}\|_2e_1\|_2.$$

This final expression can simply be seen as a minimum norm least squares problem (Barrett et al., 1994).

3.6 Generalized Minimal Residual Method

The generalized minimal residual (GMRES) method is the extended type of the MINRES method, which is also used for asymmetric matrices (Saad and Schultz, 1986). Like the MINRES method, it generates a sequence of orthogonal vectors, but in the case of asymmetric matrices this can no longer be done with short recurrences. Therefore, all orthogonal vectors that were previously obtained are stored, and thus the restarted version of this method is used (Saad and Schultz, 1986; Barrett et al., 1994).

In the CG method, the residuals form an orthogonal base for subspace

$$\mathrm{span}\{r^{(0)}, Ar^{(0)}, A^2r^{(0)}, \dots\}.$$

In the GMRES method, this base is formed by the Gram–Schmidt orthogonalization algorithm, Algorithm 3.4, (Saad, 1996). By using this algorithm for the Krylov subspace $\{A^kr^{(0)}\}$, Arnoldi orthogonalization algorithm, Algorithm 3.5, is obtained (Saad, 2011). Here the algorithm of these orthogonalizations is mentioned. It should be noted that the coefficients $(w^{(i)}, v^{(k)})$ and $\|w^{(i)}\|$ in the algorithm of the Gram–Schmidt orthogonalization are stored in a Hessenberg matrix.

Algorithm 3.4 Gram–Schmidt Orthogonalization

1: $w^{(i)} = Av^{(i)}$
2: **for** $k = 1, 2, \dots, i$ **do**
3:     $w^{(i)} = w^{(i)} - (w^{(i)}, v^{(k)})v^{(k)}$
4: **end for**
5: $v^{(i+1)} = \dfrac{w^{(i)}}{\|w^{(i)}\|}$

Algorithm 3.5 Arnoldi Orthogonalization

1: Consider vector $v^{(1)}$ with norm 1
2: **for** $j = 1, 2, \ldots, m$ **do**
3:    **for** $i = 1, 2, \ldots, j$ **do**
4:       $h_{i,j} = (Av^{(j)}, v^{(i)})$
5:       $w^{(j)} = Av^{(j)} - \sum_{i=1}^{j} h_{i,j} v^{(i)}$
6:    **end for**
7:    $h_{j+1,j} = \|w^{(j)}\|_2$
8:    $v^{(j+1)} = \dfrac{w^{(j)}}{h_{j+1,j}}$
9: **end for**

A vector of approximate solution in GMRES method can be formed as follows (Barrett et al., 1994)

$$x^{(i)} = x^{(0)} + y_1 v^{(1)} + \cdots + y_i v^{(i)}.$$

Here, the coefficients y_k are determined in such a way that the residual norm $\|b - Ax^{(i)}\|$ is minimized. In the GMRES algorithm, residual norms are calculated without the need for successive iterations. Therefore, the (costly) arithmetic operation for constituting the successive iterations is postponed until the residual norm is sufficiently diminished. The GMRES method is designed to solve asymmetric linear systems, and its most common type is based on the Gram–Schmidt process, which uses the restart algorithm to control the required memory (Saad and Schultz, 1986; Van Der Vorst, 2002). If the restart algorithm is not used, this method, like other Krylov subspace methods, converges in n iterations, but this method does not have practical value when the number of iterations rises. However, in the absence of restart, computational and memory requirements are considered as deterrent factor. In fact, the most important factor in correct application of the restart algorithm of GMRES method is choosing the restart point of iteration m. The restart algorithm is known as GMRES(m) where m is the number of iterations after which the solution process is restarted.

The GMRES method can be effectively combined (preconditioned) with other iteration based methods (Van der Vorst, 2003). The iteration steps of the GMRES method and preconditioning methods are considered as external and internal iteration steps, respectively. This hybrid method is called GMRES∗, where ∗ is the symbol of the preconditioning method (Van der Vorst and Vuik, 1994). The algorithm of GMRES(m), Algorithm 3.6, and GMRES∗, Algorithm 3.7, are as follows

---

Algorithm 3.6 GMRES(m) Method

1: `Initialization:`
2: $x^{(m)} = x^{(0)}$
3: `while` The convergence criteria is not met $x^{(0)} = x^{(m)}$ `do`
4:     $r^{(0)} = b - Ax^{(0)}$
5:     $\beta = \|r^{(0)}\|_2$
6:     $v^{(1)} = \|r^{(0)}\|/\beta$
7:     Define $\overline{H}_m = \{h_{i,j}\}_{1 \le i \le m+1, 1 \le j \le m}$ and $\overline{H}_m = 0$
8:     `for` $j = 1, 2, \ldots, m$ `do`
9:         $w^{(j)} = Av^{(j)}$
10:         `for` $i = 1, \ldots, j$ `do`
11:             $h_{i,j} = (w^{(j)}, v^{(i)})$
12:             $w^{(j)} = w^{(j)} - h_{i,j}v^{(i)}$
13:         `end for`
14:         $h_{j+1,j} = \|w^{(j)}\|_2$
15:         `if` $h_{j+1,j} = 0$ `then`
16:             $m = j$ and go to 21
17:         `else`
18:             $v^{(j+1)} = w^{(j)}/h_{j+1,j}$
19:         `end if`
20:     `end for`
21:     Compute y_m such that minimizes $\|\beta e_1 - \overline{H}_m y\|_2$
22:     $x^{(m)} = x^{(0)} + V_m y_m$
23: `end while`

3.7 Conjugate Residual Method

A new algorithm, Algorithm 3.8, can be extracted from the GMRES method, which is designed for the linear equations system with the Hermitian coefficients matrix (Saad, 1996). In this method, in addition to the fact that the residual vectors must be A-orthogonal (or conjugate), the vectors Ap_i are also orthogonal. This method has the same structure as the conjugate gradient method, with the difference that the residual vectors are conjugate with each other. The conjugate gradient and the conjugate residual methods have a similar convergence pattern, but the number of arithmetic operations and vectors required in the conjugate residual method is higher and hence the conjugate gradient method is more preferable to the conjugate residual method.

Algorithm 3.7 GMRES* Method

1: `Initialization:`
2:     $r^{(0)} = b - Ax^{(0)}$
3: `for` $i = 0, 1, 2, \ldots$ `do`
4:         `Calculate` $z^{(m)}$ `as the approximate solution of`
          $Az = r^{(i)}$ `(obtained after` m `steps of an iterative`
          `method).`
5:         $c = Az^{(m)}$
6:     `for` $k = 0, 1, \ldots, i-1$ `do`
7:         $\alpha = (c_k, c)$
8:         $c = c - \alpha c_k$
9:         $z^{(m)} = z^{(m)} - \alpha u_k$
10:     `end for`
11:         $u_i = \dfrac{z^{(m)}}{\|c\|_2}$
12:         $c_i = \dfrac{c}{\|c\|_2}$
13:         $x_{i+1} = x_i + (c_i, r_i)u_i$
14:         $r_{i+1} = r_i + (c_i, r_i)c_i$
15:     `if` r_{i+1} `is not small enough` `then`
16:             `go to 3`
17:     `else`
18:             `go to 20`
19:     `end if`
20: `end for`

3.8 Bi-Conjugate Gradient Method

The conjugate gradient method is not suitable for asymmetric systems because it is impossible to construct orthogonal residual vectors in limited iterations (Faber and Manteuffel, 1984; Voevodin, 1983). The bi-conjugate gradient method adopts another approach and replaces the orthogonal sequence of residuals with two mutually orthogonal sequences. This method is based on the asymmetric Lanczos method (Lanczos, 1952) which was introduced by Fletcher (1976), Saad and Van Der Vorst (2000). This method solves not only the equation system $Ax = b$, but also $A^T \tilde{x} = \tilde{b}$. Residual vectors and search directions corresponding to these two systems are updated as follows (Barrett et al., 1994)

$$r^{(i)} = r^{(i-1)} - \alpha_i A p^{(i)},$$
$$\tilde{r}^{(i)} = \tilde{r}^{(i-1)} - \alpha_i A^T \tilde{p}^{(i)},$$

Algorithm 3.8 Conjugate Residual Method

1: `Initialization:`

2:   $r^{(0)} = b - Ax^{(0)}$

3:   $p^{(0)} = r^{(0)}$

4: `for` $i = 0, 1, 2, \dots$ `do`

5:     $\alpha_i = \dfrac{(r^{(i)}, Ar^{(i)})}{(Ap^{(i)}, Ap^{(i)})}$

6:     $x^{(i+1)} = x^{(i)} + \alpha_i p^{(i)}$

7:     $r^{(i+1)} = r^{(i)} - \alpha_i Ap^{(i)}$

8:     `if` $r^{(i+1)}$ `is not small enough then`

9:         `go to 13`

10:     `else`

11:         `go to 16`

12:     `end if`

13:     $\beta_i = \dfrac{(r^{(i+1)}, Ar^{(i+1)})}{(r^{(i)}, Ar^{(i)})}$

14:     $p^{(i+1)} = r^{(i+1)} + \beta_i p^{(i)}$

15:     $Ap^{(i+1)} = Ar^{(i+1)} + \beta_i Ap^{(i)}$

16: `end for`

$$p^{(i)} = r^{(i-1)} + \beta_{i-1} p^{(i-1)},$$
$$\tilde{p}^{(i)} = \tilde{r}^{(i-1)} + \beta_{i-1} \tilde{p}^{(i-1)}.$$

Defining the parameters α_i and β_i as

$$\alpha_i = \frac{(r^{(i-1)}, \tilde{r}^{(i-1)})}{(Ap^{(i)}, \tilde{p}^{(i)})},$$

$$\beta_i = \frac{(r^{(i)}, \tilde{r}^{(i)})}{(r^{(i-1)}, \tilde{r}^{(i-1)})},$$

ensures the following orthogonal relations

$$(r^{(j)}, \tilde{r}^{(i)}) = (Ap^{(j)}, \tilde{p}^{(i)}) = 0,$$

if $i \neq j$.

The algorithm of the above method is presented below. There are few theoretical results about the convergence of the bi-conjugate gradient method. The bi-conjugate gradient method for a equations system with a symmetric positive-definite matrix provides similar results to the conjugate gradient method with double computational cost in each iteration. The bi-conjugate gradient method has a matrix–vector multiplication $Ap^{(i)}$ and a transpose matrix–vector multiplication $A^T \tilde{p}^{(i)}$. In some cases, it is not possible to perform the second multiplication; for example, if the matrix is not explicitly formed and stored in a compact form, then the second multiplication should be performed by calling the subroutines developed for the transpose matrix–vector multiplication. Finally, according to the presented explanations, we can present the bi-conjugate gradient algorithm, Algorithm 3.9, as follows.

Algorithm 3.9 Bi-Conjugate Gradient Method

1: **Initialization:**

2: $\quad r^{(0)} = b - Ax^{(0)}$

3: $\quad$ Choose an arbitrary $\tilde{r}^{(0)}$ such that $(r^{(0)}, \tilde{r}^{(0)}) \neq 0$, usually $\tilde{r}^{(0)} = r^{(0)}$.

4: $\quad p^{(0)} = r^{(0)}$

5: $\quad \tilde{p}^{(0)} = \tilde{r}^{(0)}$

6: **for** $i = 0, 1, 2, \ldots$ **do**

7: $\quad\quad \alpha_i = \dfrac{(r^{(i)}, \tilde{r}^{(i)})}{(Ap^{(i)}, \tilde{r}^{(i)})}$

8: $\quad\quad x^{(i+1)} = x^{(i)} + \alpha_i p^{(i)}$

9: $\quad\quad r^{(i+1)} = r^{(i)} - \alpha_i Ap^{(i)}$

10: $\quad$ **if** $r^{(i+1)}$ is not small enough **then**

11: $\quad\quad$ go to 15

12: $\quad$ **else**

13: $\quad\quad$ go to 19

14: $\quad$ **end if**

15: $\quad\quad \tilde{r}^{(i+1)} = \tilde{r}^{(i)} - \alpha_i A^T \tilde{p}^{(i)}$

16: $\quad\quad \beta_i = \dfrac{(r^{(i+1)}, \tilde{r}^{(i+1)})}{(r^{(i)}, \tilde{r}^{(i)})}$

17: $\quad\quad p^{(i+1)} = r^{(i+1)} + \beta_i p^{(i)}$

18: $\quad\quad \tilde{p}^{(i+1)} = \tilde{r}^{(i+1)} + \beta_i \tilde{p}^{(i)}$

19: **end for**

3.9 Transpose-Free Methods

In each iteration step of bi-conjugate gradient algorithm, two operations of matrix–vector multiplication are performed, one with the main matrix and the other with its corresponding transpose matrix. However, it can be seen that vectors $\tilde{p}^{(i)}$ and $\tilde{r}^{(i)}$, which are updated through transpose matrix in each iteration, do not have a direct role in solution, and are used only to obtain the scalar quantities, α_i and β_i, required by the bi-conjugate gradient algorithm (Saad, 1996). The question that arises here is whether it is possible to avoid using transpose matrix? One of the motivations that led to this question is that in some cases the matrix A is not explicitly available and can only be obtained by estimation and transpose matrix is usually not available.

3.9.1 Conjugate Gradient Squared Method

The Conjugate Gradient Squared method was developed by Sonneveld (1989). In this method, the use of the transpose matrix is avoided and for the same computational cost, leads to faster convergence. The main idea of this method is based on the observations that will be discussed below. In the bi-conjugate gradient method, the residual vector in ith iteration step can be represented as

$$r^{(i)} = \phi_i(A)r^{(0)},$$

where ϕ_i is a polynomial of degree i and $\phi_i(0) = 1$. It can also be shown that

$$p^{(i)} = \pi_i(A)r^{(0)},$$

where π_i is a polynomial of degree i. In the bi-conjugate gradient method, the vectors $\tilde{r}^{(i)}$ and $\tilde{p}^{(i)}$ are updated with recurrence relations (similar to those for vectors $r^{(i)}$ and $p^{(i)}$), in which A is replaced by A^T; so

$$\tilde{r}^{(i)} = \phi_i(A^T)\tilde{r}^{(0)},$$

$$\tilde{p}^{(i)} = \pi_i(A^T)\tilde{r}^{(0)}.$$

Also, rewriting the recurrence relation for the quantity α_i in the bi-conjugate gradient method gives

$$\alpha_i = \frac{(\phi_i(A)r^{(0)}, \phi_i(A^T)\tilde{r}^{(0)})}{(A\pi_i(A)r^{(0)}, \pi_i(A^T)\tilde{r}^{(0)})} = \frac{(\phi_i^2(A)r^{(0)}, \tilde{r}^{(0)})}{(A\pi_i^2(A)r^{(0)}, \tilde{r}^{(0)})}.$$

If we can write a recurrence relation for the vectors $\phi_i^2(A)r^{(0)}$ and $\pi_i^2(A)r^{(0)}$, calculating α_i and β_i will not encounter any problems. Derivation of this method relies on simple algebraic relations. To form the recurrence relations for squared polynomials, first, the recurrence relations for $\phi_i(A)$ and $\pi_i(A)$ should be written, which are

$$\phi_{i+1}(t) = \phi_i(t) - \alpha_i t \pi_i(t), \tag{3.15}$$

$$\pi_{i+1}(t) = \phi_{i+1}(t) + \beta_i \pi_i(t). \tag{3.16}$$

If the above relations are squared, one can directly find

$$\phi_{i+1}^2(t) = \phi_i^2(t) - 2\alpha_i t \pi_i(t)\phi_i(t) + \alpha_i^2 t^2 \pi_i^2(t),$$

$$\pi_{i+1}^2(t) = \phi_{i+1}^2(t) + 2\beta_i \phi_{i+1}(t)\pi_i(t) + \beta_i^2 \pi_i(t)^2.$$

Because of the cross terms $\pi_i(t)\phi_i(t)$ and $\phi_{i+1}(t)\pi_i(t)$ on the right-hand side of the above relations, the recurrence system cannot be formed. Therefore, we must obtain the recurrence relation for the terms $\pi_i(t)\phi_i(t)$ and $\phi_{i+1}(t)\pi_i(t)$. Using equations Equation (3.15) and Equation (3.16) we have

$$\phi_i(t)\pi_i(t) = \phi_i(t)(\phi_i(t) + \beta_{i-1}\pi_{i-1}(t)) = \phi_i^2(t) + \beta_{i-1}\phi_i(t)\pi_{i-1}(t), \tag{3.17}$$

$$\phi_{i+1}(t)\pi_i(t) = \pi_i(t)(\phi_i(t) - \alpha_i t \pi_i(t)) = \phi_i(t)\pi_i(t) - \alpha_i t \pi_i^2(t). \tag{3.18}$$

Inserting the Equation (3.17) into Equation (3.18) and compiling the previous relations, results in the following recurrence relations

$$\phi_{i+1}^2(t) = \phi_i^2(t) - \alpha_i t(2\phi_i^2(t) + 2\beta_{i-1}\phi_i(t)\pi_{i-1}(t) - \alpha_i t\pi_i^2(t)),$$
$$\phi_{i+1}(t)\pi_i(t) = \phi_i^2(t) + \beta_{i-1}\phi_i(t)\pi_{i-1}(t) - \alpha_i t\pi_i^2(t),$$
$$\pi_{i+1}^2(t) = \phi_{i+1}^2(t) + 2\beta_i\phi_{i+1}(t)\pi_i(t) + \beta_i^2\pi_i^2(t).$$

The above-mentioned recurrence relations constitute the main part of the conjugate gradient squared method. One can define

$$r^{(i)} = \phi_i^2(A)r^{(0)},$$
$$p^{(i)} = \pi_i^2(A)r^{(0)},$$
$$q^{(i)} = \phi_{i+1}(A)\pi_i(A)r^{(0)},$$

and the polynomials recurrence relations can be rewritten as

$$r^{(i+1)} = r^{(i)} - \alpha_i A(2r^{(i)} + 2\beta_{i-1}q^{(i-1)} - \alpha_i Ap^{(i)}),$$
$$q^{(i)} = r^{(i)} + \beta_{i-1}q^{(i-1)} - \alpha_i Ap^{(i)},$$
$$p^{(i+1)} = r^{(i+1)} + 2\beta_i q^{(i)} + \beta_i^2 p^{(i)}.$$

We can simply define the vector $d^{(i)}$ as follows

$$d^{(i)} = 2r^{(i)} + 2\beta_{i-1}q^{(i-1)} - \alpha_i Ap^{(i)}.$$

By compiling the relations that have been described so far, the consecutive operations used to compute the approximate answer are

- $\alpha_i = (r^{(i)}, \tilde{r}^{(0)})/(Ap^{(i)}, \tilde{r}^{(0)})$
- $d^{(i)} = 2r^{(i)} + 2\beta_{i-1}q^{(i-1)} - \alpha_i Ap^{(i)}$
- $q^{(i)} = r^{(i)} + \beta_{i-1}q^{(i-1)} - \alpha_i Ap^{(i)}$
- $x^{(i+1)} = x^{(i)} + \alpha_i d^{(i)}$
- $r^{(i+1)} = r^{(i)} - \alpha_i Ad^{(i)}$
- $\beta_i = (r^{(i+1)}, \tilde{r}^{(0)})/(r^{(i)}, \tilde{r}^{(0)})$
- $p^{(i+1)} = r^{(i+1)} + \beta_i(2q^{(i)} + \beta_i p^{(i)})$

The initial values in this algorithm are

$$r^{(0)} = b - Ax^{(0)},$$
$$p^{(0)} = r^{(0)},$$
$$q^{(0)} = 0,$$
$$\beta_0 = 0.$$

By defining the auxiliary vector $u^{(i)} = r^{(i)} + \beta_{i-1}q^{(i-1)}$, the solution algorithm is simplified and leads to the following relations

$$d^{(i)} = u^{(i)} + q^{(i)},$$
$$q^{(i)} = u^{(i)} - \alpha_i Ap^{(i)},$$
$$p^{(i+1)} = u^{(i+1)} + \beta_i(q^{(i)} + \beta_i p^{(i)}).$$

It should be noted that according to the presented relations, obtaining the vector $d^{(i)}$ does not seem to be more than necessary, and the final algorithm of the conjugate gradient squared method is expressed in Algorithm 3.10. As seen in the conjugate gradient squared algorithm, the matrix–vector multiplication is independent of the transpose matrix and only the original matrix is used. Thus, at each iteration step, there are two matrix–vector multiplications (with the original matrix A) and it is expected that the convergence rate of this method will be twice as fast as the bi-conjugate gradient method. The conjugate gradient squared algorithm works well in many cases, but in some systems meets some difficulties. Since the polynomials formed in the solution process are squared, the round-off errors in this method may be more acute than the bi-conjugate gradient method. For example, extreme changes in residual vectors often lead to a reduction in the accuracy of the residual norms, calculated from the results of line eight in Algorithm 3.10.

Algorithm 3.10 Conjugate Gradient Squared Method

1: **Initialization:**

2:     $r^{(0)} = b - Ax^{(0)}$

3:     Choose an arbitrary $\tilde{r}^{(0)}$ such that $(r^{(0)}, \tilde{r}^{(0)}) \neq 0$, usually $\tilde{r}^{(0)} = r^{(0)}$.

4:     $u^{(0)} = r^{(0)}$

5:     $p^{(0)} = r^{(0)}$

6: **for** $i = 0, 1, 2, \ldots$ **do**

7:     $\alpha_i = \dfrac{(r^{(i)}, \tilde{r}^{(0)})}{(Ap^{(i)}, \tilde{r}^{(0)})}$

8:     $q^{(i)} = u^{(i)} - \alpha_i Ap^{(i)}$

9:     $x^{(i+1)} = x^{(i)} + \alpha_i(u^{(i)} + q^{(i)})$

10:     $r^{(i+1)} = r^{(i)} - \alpha_i A(u^{(i)} + q^{(i)})$

11:     **if** $r^{(i+1)}$ is not small enough **then**

12:         go to 17

13:     **else**

14:         go to 20

15:     **end if**

16:     $\beta_i = \dfrac{(r^{(i+1)}, \tilde{r}^{(0)})}{(r^{(i)}, \tilde{r}^{(0)})}$

17:     $u^{(i+1)} = r^{(i+1)} + \beta_i q^{(i)}$

18:     $p^{(i+1)} = u^{(i+1)} + \beta_i(q^{(i)} + \beta_i p^{(i)})$

19: **end for**

3.9.2 Bi-Conjugate Gradient Stabilized Method

The conjugate gradient squared method is based on squaring the polynomial of the residuals, and in cases where the convergence process is irregular, it can essentially lead to accumulation of round-off errors and even memory overflow (Saad, 1996).

A bi-conjugate gradient stabilized method is another kind of conjugate gradient squared method that remedies this defect. In the bi-conjugate gradient stabilized method the residual vectors are formed based on the following relation

$$r^{(i)} = \psi_i(A)\phi_i(A)r^{(0)},$$

Where $\phi_i(A)$, as mentioned in Section 3.9.1, is a polynomial of residuals and is related to the bi-conjugate gradient algorithm and $\psi_i(A)$ is a new polynomial that is defined recursively in each iteration and its function is to stabilize or smooth the convergence behavior of the previous method (Barrett et al., 1994; Saad, 1996; Van der Vorst, 2003). This smoothing polynomial can be defined by a simple recurrence relation, i.e.

$$\psi_{i+1}(t) = (1 - \omega_i t)\psi_i.$$

Here the constant ω_i must be determined. The extraction technique of proper recurrence relations for this method is similar to that of the conjugate gradient squared method. First, by ignoring the scalar coefficients, we obtain a recurrence relation for the polynomial $\psi_{i+1}\phi_{i+1}$, i.e.

$$\psi_{i+1}\phi_{i+1} = (1 - \omega_i t)\psi_i\phi_{i+1} = (1 - \omega_i t)(\psi_i\phi_i - \alpha_i t\psi_i\pi_i),$$

where a recurrence relation for $\psi_i\pi_i$ should also be found. Hence,

$$\psi_i\pi_i = \psi_i(\phi_i + \beta_{i-1}\pi_{i-1}) = \psi_i\phi_i + \beta_{i-1}(1 - \omega_{i-1}t)\psi_{i-1}\pi_{i-1}.$$

Given the above recurrence relations and knowing that

$$r^{(i)} = \phi_i(A)\psi_i(A)r^{(0)},$$

$$p^{(i)} = \psi_i(A)\pi_i(A)r^{(0)},$$

the recurrence relations for these two vectors can be written as follows

$$r^{(i+1)} = (I - \omega_i A)(r^{(i)} - \alpha_i A p^{(i)}), \tag{3.19}$$

$$p^{(i+1)} = r^{(i+1)} + \beta_i(I - \omega_i A)p^{(i)}.$$

Calculation of scalar coefficients is an important part of the algorithm. In the bi-conjugate gradient method $\beta_i = \rho_{i+1}/\rho_i$ and

$$\rho_i = (\phi_i(A)r^{(0)}, \phi_i(A^T)\tilde{r}^{(0)}) = (\phi_i^2(A)r^{(0)}, \tilde{r}^{(0)}).$$

Since none of the vectors $\phi_i(A)r^{(0)}$, $\phi_i(A^T)\tilde{r}^{(0)}$ and $\phi_i^2(A)r^{(0)}$ are available, ρ_i cannot be calculated from the above relation. However, ρ_i can be somehow related to following scalar quantity

$$\tilde{\rho}_i = (\phi_i(A)r^{(0)}, \psi_i(A^T)\tilde{r}^{(0)})$$

$$= (\psi_i(A)\phi_i(A)r^{(0)}, \tilde{r}^{(0)})$$

$$= (r^{(i)}, \tilde{r}^{(0)}).$$

By expanding $\psi_i(A^T)$ into power bases, one can find that

$$\tilde{\rho}_i = (\phi_i(A)r^{(0)}, \eta_1^{(i)}(A^T)^j\tilde{r}^{(0)} + \eta_2^{(i)}(A^T)^{j-1}\tilde{r}^{(0)} + \cdots).$$

Since $\phi_i(A)r^{(0)}$ is orthogonal to all vectors $(A^T)^k \tilde{r}^{(0)}(k < j)$, only the first term of the above power expansion has a effective role. Therefore, if $\gamma_1^{(i)}$ is the first coefficient of the polynomial $\phi_i(t)$, we can write

$$\tilde{\rho}_i = (\phi_i(A)r^{(0)}, \frac{\eta_1^{(i)}}{\gamma_1^{(i)}} \phi_i(A^T)\tilde{r}^{(0)}) = \frac{\eta_1^{(i)}}{\gamma_1^{(i)}} \rho_i.$$

By evaluating the recurrence relations ϕ_{i+1} and ψ_{i+1}, we can obtain a recurrence relation for the first coefficients of these polynomials. Thus

$$\eta_1^{(i+1)} = -\omega_i \eta_1^{(i)},$$

$$\gamma_1^{(i+1)} = -\alpha_i \gamma_1^{(i)},$$

and as a result

$$\frac{\tilde{\rho}_{i+1}}{\tilde{\rho}_i} = (\frac{\omega_i}{\alpha_i})(\frac{\rho_{i+1}}{\rho_i}),$$

which results in the following relation for β_i, i.e.

$$\beta_i = (\frac{\tilde{\rho}_{i+1}}{\tilde{\rho}_i})(\frac{\alpha_i}{\omega_i}).$$

Similarly, a simple recurrence relation for α_i can be obtained. As mentioned in Subsection 3.9.1,

$$\alpha_i = \frac{(\phi_i(A)r^{(0)}, \phi_i(A^T)\tilde{r}^{(0)})}{(A\pi_i(A)r^{(0)}, \pi_i(A^T)\tilde{r}^{(0)})},$$

and as mentioned above, the polynomials derived from the inner products of the numerator and the denominator can be replaced by their first terms. However, here the coefficients of the first terms of $\phi_i(A^T)\tilde{r}^{(0)}$ and $\pi_i(A^T)\tilde{r}^{(0)}$ are similar and, therefore,

$$\alpha_i = \frac{(\phi_i(A)r^{(0)}, \phi_i(A^T)\tilde{r}^{(0)})}{(A\pi_i(A)r^{(0)}, \phi_i(A^T)\tilde{r}^{(0)})}$$

$$= \frac{(\phi_i(A)r^{(0)}, \psi_i(A^T)\tilde{r}^{(0)})}{(A\pi_i(A)r^{(0)}, \psi_i(A^T)\tilde{r}^{(0)})}$$

$$= \frac{(\psi_i(A)\phi_i(A)r^{(0)}, \tilde{r}^{(0)})}{(A\psi_i(A)\pi_i(A)r^{(0)}, \tilde{r}^{(0)})}.$$

Since, $p^{(i)} = \psi_i(A)\pi_i(A)$, the former equation can be presented in a simpler way, i.e.

$$\alpha_i = \frac{\tilde{\rho}_i}{(Ap^{(i)}, \tilde{r}^{(0)})}.$$

In the next step, coefficient ω_i, should be determined. This coefficient is obtained in such a way to minimize the Euclidean norm of the vector $(I - \omega_i A)\psi_i(A)\phi_{i+1}(A)r^{(0)}$. Hence, Equation (3.19) can be rewritten as

$$r^{(i+1)} = (I - \omega_i A)s^{(i)},$$

where

$$s^{(i)} \equiv r^{(i)} - \alpha_i A p^{(i)}.$$

Thus, an optimum value of the coefficient ω_i is obtained as follows

$$\omega_i = \frac{(As^{(i)}, s^{(i)})}{(As^{(i)}, As^{(i)})}.$$

Eventually, a relation for updating the approximate solution should be extracted. Therefore, Equation (3.19) should be rewritten as follows

$$r^{(i+1)} = s^{(i)} - \omega_i As^{(i)} = r^{(i)} - \alpha_i A p^{(i)} - \omega_i As^{(i)},$$

which results in the following recurrence relation for updating the approximate solution

$$x^{(i+1)} = x^{(i)} + \alpha_i p^{(i)} + \omega_i s^{(i)}.$$

Here the final algorithm, Algorithm 3.11, of the bi-conjugate gradient stabilized method is presented

---

Algorithm 3.11 Bi-Conjugate Gradient Stabilized Method

1: **Initialization:**
2: $\quad r^{(0)} = b - Ax^{(0)}$
3: $\quad$ Choose an arbitrary $\tilde{r}^{(0)}$ such that $(r^{(0)}, \tilde{r}^{(0)}) \neq 0$, usually $\tilde{r}^{(0)} = r^{(0)}$.
4: $\quad p^{(0)} = r^{(0)}$
5: **for** $i = 0, 1, 2, \ldots$ **do**
6: $\quad\quad \alpha_i = \frac{(r^{(i)}, \tilde{r}^{(0)})}{(Ap^{(i)}, \tilde{r}^{(0)})}$
7: $\quad\quad s^{(i)} = r^{(i)} - \alpha_i A p^{(i)}$
8: $\quad\quad \omega_i = \frac{(As^{(i)}, s^{(i)})}{(As^{(i)}, As^{(i)})}$
9: $\quad\quad x^{(i+1)} = x^{(i)} + \alpha_i p^{(i)} + \omega_i s^{(i)}$
10: $\quad\quad r^{(i+1)} = s^{(i)} - \omega_i As^{(i)}$
11: $\quad\quad$ **if** $r^{(i+1)}$ is not small enough **then**
12: $\quad\quad\quad$ go to 17
13: $\quad\quad$ **else**
14: $\quad\quad\quad$ go to 20
15: $\quad\quad$ **end if**
16: $\quad\quad \beta_i = \frac{(r^{(i+1)}, \tilde{r}^{(0)})}{(r^{(i)}, \tilde{r}^{(0)})} \times \frac{\alpha_i}{\omega_i}$
17: $\quad\quad p^{(i+1)} = r^{(i+1)} + \beta_i(p^{(i)} - \omega_i A p^{(i)})$
18: **end for**

Exercises

3.1 Consider a 2×2 system of equations $Mx = b$ where matrix M is as follows

$$\begin{pmatrix} 1 & \alpha \\ \alpha & 1 \end{pmatrix}.$$

The following table shows the variations of the L^2-norm based condition number of the matrix A according to α

α	1.1	1.2	1.3	1.4	1.5	2.0	5.0	10.0	100.0	1000.0
κ_2	21.00	11.00	7.66	6.00	5.00	3.00	1.50	1.22	1.02	1.00

(a) Apply the steepest descent Algorithm 3.2 and solve the foregoing system of equations for an arbitrary right-hand side vector with various values of α, using hand calculations or rudimentary spreadsheet programs. It is obvious that the hand calculations are cumbersome, specially in the cases with greater condition numbers; so you can truncate the calculations when you understand the convergence or divergence trend of the method.

(b) Report the convergence or divergence of the solution for each case.

(c) Qualitatively, compare the rate of approaching the exact solution when the α, and correspondingly κ_2, changes.

(d) Run the program of Appendix F for the current exercise and compare the results with the results of the hand calculations.

3.2 Apply the CG method Algorithm 3.3 and response to the demands (a) to (c) of Exercise 1.

3.3 Consider a 2×2 system of equations $Nx = b$ where matrix N is as follows

$$\begin{pmatrix} 1 & \beta \\ 1 & 1 \end{pmatrix}.$$

The following table shows the variations of the L^2-norm based condition number of the matrix A according to β

β	1.1	1.2	1.3	1.4	1.5	2.0	5.0	10.0	100.0	1000.0
κ_2	42.07	22.15	15.56	12.31	10.40	6.85	6.85	11.35	101.03	1001.00

(a) Apply the BiCG method Algorithm 3.9 and solve the foregoing system of equations for an arbitrary right-hand side vector with various values of β, using hand calculations or rudimentary spreadsheet programs.

(b) Qualitatively, compare the rate of approaching the exact solution when the β, and correspondingly κ_2, changes.

(c) Apply the CGS method Algorithm 3.10 and the BiCGSTAB method Algorithm 3.11 and response to demands of two previous parts.

3.4 Analyze the arithmetic cost, i.e., the number of operations, of the CG method Algorithm 3.3 and CR method Algorithm 3.8. Similarly analyze the memory requirements of both algorithms.

3.5 If A is a symmetric positive-definite matrix and $\hat{x}$ is the exact solution of the linear system $Ax = b$, for the conjugate gradient method,

(a) Prove that

$$\|x^{(i)} - \hat{x}\|_A \leq 2\zeta^i \|x^{(0)} - \hat{x}\|_A,$$

where $\zeta = (\sqrt{\kappa} - 1)/(\sqrt{\kappa} + 1)$ and $\|y\|_A^2 = (y, Ay)$.

(b) Prove that the number of iterations needed to achieve a relative reduction in error is proportional to $\sqrt{\kappa}$.

3.6 Apply the GMRES(m) method Algorithm 3.6, for an arbitrary value of m, response to the demands of parts (a) and (b) of Exercise 3.

4

Numerical Analysis of Krylov Subspace Methods

4.1 Numerical Solution of Linear Systems

In this section we study the performance of the Krylov subspace methods in solving various problems and discuss the strength and weaknesses of these methods. As mentioned in Section 2.4, here we want to use our provided sparse matrices to construct linear systems. Then these systems are numerically solved by developed codes based on Krylov subspace solvers as given in algorithms of Chapter 3. The codes are written in the FORTRAN 90 programing language and run on the Lahey, Developer Studio and Code Blocks compiler environments. Computations are performed using an Intel(R) Core(TM) i3 processor with 3Mb cache and clock speed of 2.13 GHz.

At first glance, the goal is to solve the system $Ax = b$. In this system the coefficient matrix A is chosen from the benchmark collection of Chapter 2. It should be also noted that vector b is defined as

$$b = Ae,$$

where $e = [1, 1, \ldots, 1]^T$. In all problems that will be investigated, the initial guess is considered zero and convergence criteria is satisfied when

$$\frac{\|r\|_2}{N} < 10^{-10},$$

where $\|r\|_2$ represents the second standard norm or Euclidean norm of residual vector r and N is the length of unknown vector or rank of matrix A.

Here, the sparse matrices that play the role of coefficient matrix, generally arise from discretization of equations which model the vast range of phenomena in nature and industry. In this chapter, matrices with different characteristics are chosen to provide a basis in which analysis and comparison of methods can be easily and fully covered. Selected benchmark matrices can be categorized in three main groups, symmetric positive-definite, symmetric indefinite and asymmetric matrices. In this section, we will consider some of the benchmark matrices from

Krylov Subspace Methods with Application in Incompressible Fluid Flow Solvers, First Edition. Iman Farahbakhsh.
© 2020 John Wiley & Sons Ltd. Published 2020 by John Wiley & Sons Ltd.
Companion Website: www.wiley.com/go/Farahbakhs/KrylovSubspaceMethods

each category for constituting a system of equations and then with applying the different Krylov subspace methods, compare the convergence behavior, processing time, reduction factor, and error.

4.1.1 Solution of Symmetric Positive-Definite Systems

Here we choose three types of symmetric positive-definite matrices, NOS4, NOS7 and GR3030, for constituting the system of equations and comparing the methods. Matrix NOS7 results from finite difference approximation of the diffusion equation with variable diffusivity in a unit cube with Dirichlet boundary conditions; NOS4 arises from the finite element approximation of a beam structure and GR3030 is the result of the finite difference approximation of Laplace's operator. The convergence behavior of five methods CG, BiCG, CR, CGS and BiCGSTAB in the case of the system of equations with GR3030 coefficient matrix, hereafter called the case of GR3030 and similarly for other matrices, is illustrated in Figure 4.1. As pointed out in Section 3.8, the convergence behavior of the BiCG method in solution of systems with symmetric coefficient matrix is similar to CG method with double arithmetic

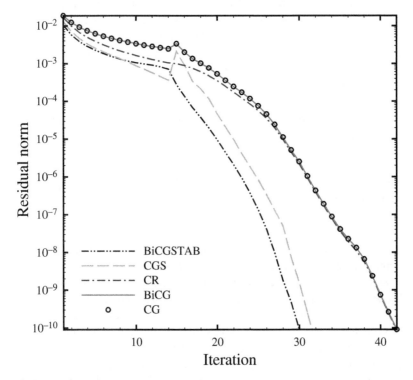

Figure 4.1 Convergence behavior of Krylov subspace methods in solution of a linear system in the case of GR3030.

operations (Barrett et al., 1994); therefore, this method is not recommended for symmetric positive-definite systems. Another point of Figure 4.1 is that the convergence type is superlinear (see Section 3.4.3). As illustrated in Figure 4.1, the convergence slope of CG is increased considerably after 20 iterations; this pattern can be also observed in other methods (Van der Vorst, 2003). The convergence rate of CG depends on the matrix spectrum characteristics and the condition number plays an important role in it (see Section 3.4.3).

While BiCGSTAB and CGS methods are developed for solving asymmetric systems, they prove to be efficient in many cases of symmetric linear systems. As can be seen in Figure 4.1, the BiCGSTAB method converges with the least number of iterations. It should be also noted that in this special system, convergence of all methods has smooth behaviors (especially for CR), such that there is not a considerable difference between the convergence trace of BiCGSTAB and CGS; but in the case of asymmetric systems, BiCGSTAB has a smoother convergence trace compared to CGS.

Figure 4.2 shows the convergence behavior of CG, BiCG, CR and BiCGSTAB methods when matrix NOS4 is considered as the coefficient matrix. The CGS

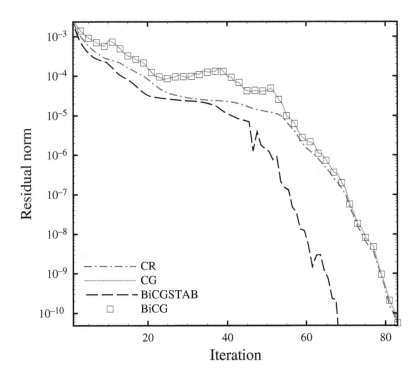

Figure 4.2 Convergence behavior of Krylov subspace methods in solution of a linear system in the case of NOS4.

method is not included in this figure, since this method is unable to minimize the second norm of residual vector below 10^{-10} and consequently diverges. As can be seen in Figure 4.2, the convergence process of the CG and BiCG methods is completely identical, and the superlinear convergence phenomenon is also formed after approximately 50 iterations.

Another parameter that can be used for analyzing the methods is reduction factor which is defined as $\|r^{(n)}\|_2 / \|r^{(n-1)}\|_2$. Averaging the reduction factor in consecutive iterations results in a mean reduction factor which is an appropriate criteria to assess the rate of error reduction and convergence smoothness in different methods. The mean reduction factors of BiCGSTAB and CGS methods in the case of GR3030 are 0.573 and 0.705, respectively. Intense fluctuations in convergence behavior leads to increase in mean reduction factor. Figure 4.3 shows the variations of the reduction factor in the case GR3030. The mean reduction factor for BiCG and CG methods is 0.658 while for CR is 0.651. It is not possible to present certain comments on this special case since the processor time was too short for all methods. Figure 4.4 shows the reduction factor variations versus iteration for converged methods in the case of NOS4. The CR method has the lowest mean

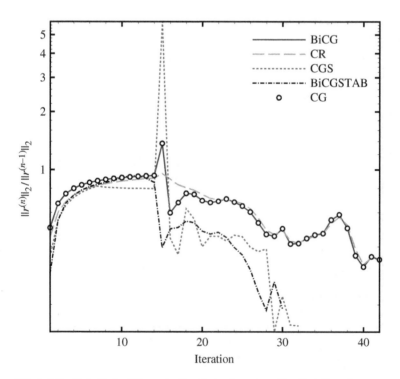

Figure 4.3 Variations of the reduction factor in consecutive iterations in convergence process of various methods for the case of GR3030.

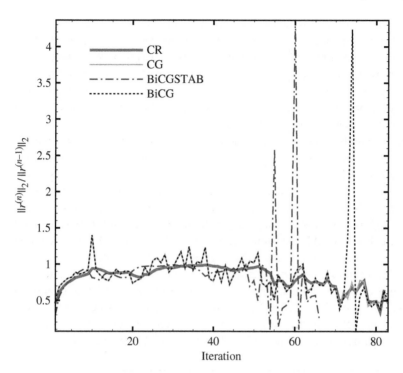

Figure 4.4 Variations of the reduction factor in consecutive iterations in the convergence process of various methods for the case of NOS4.

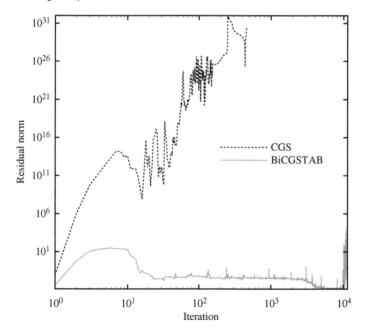

Figure 4.5 Divergence of CGS and BiCGSTAB methods in the case of NOS7.

reduction factor of 0.820 which is associated with the smooth nature of its convergence. The CG method has a few more fluctuations than the CR method which results in the mean reduction factor of 0.826. The BiCG and CG methods have similar convergence traces up to approximately 70 iterations; after that the BiCG method convergence trace deviates from the CG method due to accumulation of round-off errors. Intense fluctuations of BiCG and BiCGSTAB methods increase the mean reduction factor up to 0.869 and 0.872, respectively.

Consider another system of equations with the symmetric positive-definite coefficient matrix NOS7. In this case, BiCGSTAB and CGS methods diverge, see Figure 4.5. This divergence is sourced by the accumulation of errors in the multiplication of the residual vector polynomials, see Sections 3.9.1 and 3.9.2. However, the BiCG method converges with a convergence behavior similar to CG. Even though rank of matrix, and as a result, there are fewer unknowns than the latter case, but since the condition number is big (see Table 2.1), then required iterations for convergence is increased. Figure 4.6 shows the intense fluctuations of the CG and BiCG methods. The mean reduction factor in the CG and BiCG methods is 2.328 and in the CR method is 0.996. Although, the BiCG is expected to behave in a similar way to CG, theoretically, but for problems with higher sensitivity in coefficient matrix, i.e., higher condition number, their convergence behavior will deviate from each other. This is caused by accumulated errors in

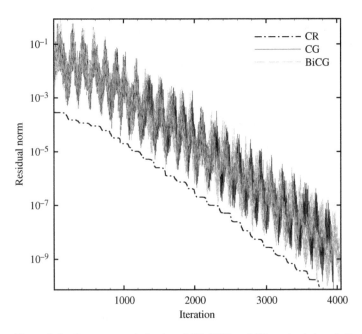

Figure 4.6 Convergence behavior of CG, BiCG and CR methods in solution of a linear system in the case of NOS7.

consecutive iterations. This issue is well depicted in Figure 4.7. Table 4.1 presents maximum error and processing time of each method. It should be noted that maximum error is calculated with regard to the exact solution mentioned at the beginning of the chapter.

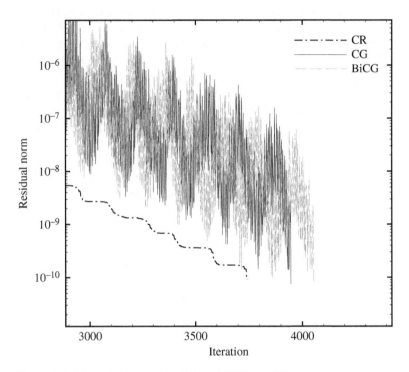

Figure 4.7 The residual norm deviation of BiCG from CG.

Table 4.1 Maximum exact error and processing time in the solution of systems with coefficient matrices GR3030, NOS7, and NOS4.

	GR3030		NOS7		NOS4	
	Maximum Error	Processing Time [sec]	Maximum Error	Processing Time [sec]	Maximum Error	Processing Time [sec]
CG	1.61e-09	0.01	1.85e-07	0.90	8.46e-09	0[a)]
BiCG	1.61e-09	0.01	8.15e-08	1.06	8.46e-09	0
CGS	8.97e-10	0.01	-	-	-	-
BiCGSTAB	7.65e-08	0.01	-	-	3.86e-07	0
CR	1.94e-09	0.01	8.38e-07	1.60	1.66e-08	0.01

a) Processing time less than 0.0156 sec is presented by 0

4.1.2 Solution of Asymmetric Systems

Here, the asymmetric matrices ORSREG1, SHERMAN4, ORSIRR2 and ORSIRR1 are selected to form the system of equations. First let us look at how the methods work to solve the ORSREG1 equation system. It will not be surprising if CG and CR solvers diverge in the solution process of these four cases. The reason, as pointed out in Section 3.8, is that they are not able to minimize residual vectors with short recursive relations and we have to resort to other methods. Convergence behavior of BiCG, CGS and BiCGSTAB methods for asymmetric matrix ORSREG1 is given in Figure 4.8. As can be seen in this figure, the CGS method has no regular convergence and its mean reduction factor is 564 978.76 which is most undesirable. Applying the BiCG and the BiCGSTAB methods decreases the mean reduction factor to 13.166 and 2.997, respectively. As shown in Figure 4.8, the convergence of the BiCGSTAB method depicts less fluctuations compared to that of the CGS and the overall trend is more regular. This issue also holds in the case of system of equations with the matrix SHERMAN4, see Figure 4.9. The mean reduction factor of the CGS, BiCG, BiCGSTAB, CG and CR methods in the case of SHERMAN4

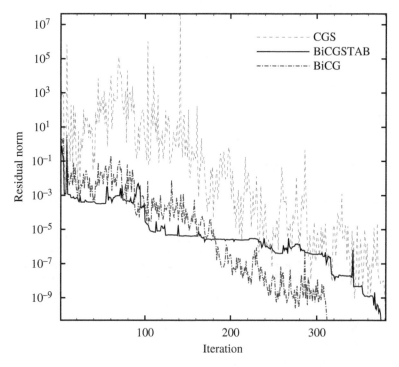

Figure 4.8 The convergence behavior of BiCG, CGS and BiCGSTAB methods in solution of a linear system with asymmetric coefficient matrix ORSREG1.

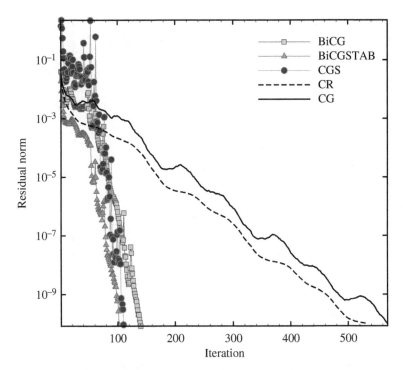

Figure 4.9 The convergence behavior of BiCG, CGS and BiCGSTAB methods in solution of a linear system with asymmetric coefficient matrix SHERMAN4.

are 71.197, 1.008, 0.886, 0.968 09 and 0.964 95, respectively. Here, the CG and CR methods also converge, but the number of iterations required for their convergence is several times greater than that of the other methods. Severe variations of the second norm of the residual vectors (Figure 4.10) can sometimes lead to divergence of the CGS method and even if it converges, the maximum error may have the same order as the solution. The CGS method is also sensitive to the choice of the initial guess, so that if the initial guess is accidentally close to the solution of the system, it may diverge (Barrett et al., 1994). According to the foregoing discussions, more caution should be taken in choosing the CGS method for solving the asymmetric system of equations.

Table 4.2 shows the convergence CPU-time as well as the maximum error in solving the asymmetric system of equations by CGS, BiCG, and BiCGSTAB methods. As indicated in Table 4.2, the CGS method diverges in ORSIRR1 and ORSIRR2 cases. It is also found that from the view point of the convergence CPU-time and the maximum error, the BiCG method is generally the most suitable method, and preferred in problems where making the transpose matrix is feasible.

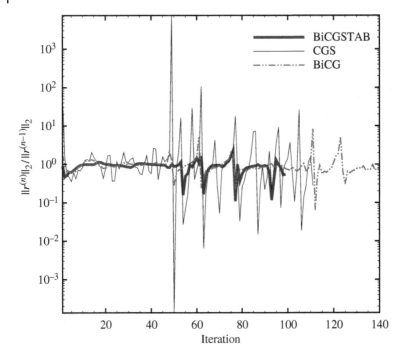

Figure 4.10 Variations of the reduction factor in consecutive iterations of BiCG, CGS, and BiCGSTAB methods in the solution of a linear system with asymmetric coefficient matrix SHERMAN4.

Table 4.2 Maximum exact error and processing time in the solution of systems with coefficient matrices SHERMAN4, ORSREG1, ORSIRR1, and ORSIRR2.

	ORSREG1		SHERMAN4	
	Maximum Error	Processing Time [sec]	Maximum Error	Processing Time [sec]
BiCG	6.68e-11	0.24	3.12e-09	0.06
CGS	2.75e-09	0.49	4.97e-10	0.07
BiCGSTAB	8.22e-10	0.48	3.20e-08	0.06

	ORSIRR2		ORSIRR1	
	Maximum Error	Processing Time [sec]	Maximum Error	Processing Time [sec]
BiCG	3.26e-11	0.34	6.06e-11	0.54
CGS	-	-	-	-
BiCGSTAB	3.48e-10	0.82	1.38e-09	1.06

4.1.3   Solution of Symmetric Indefinite Systems

In this subsection, we consider two set of equations with symmetric indefinite matrices BCSSTK22 and PLAT362. Although the CG method is efficient in the case of symmetric positive-definite matrices, it should not be ignored that definiteness of coefficient matrix ensures the existence of minimum value for the quadratic form of the coefficient matrix (McDonough, 2008) (see Section 3.4.2). Therefore, in the case of indefinite matrices it is not possible to certainly remark on convergence or divergence of the CG method. Figure 4.11 shows the convergence process of the CR, BiCGSTAB, BiCG, and CG methods in solving the system of equations with matrix BCSSTK22. As expected, convergence behavior of the BiCG and CG methods are the same but the convergence CPU-time of the CG is half of BiCG. Among the implemented methods, CR has the least mean reduction factor, i.e., 0.934, due to its smooth convergence. Mean reduction factor for BiCGSTAB, BiCG and CG methods are 1.115, 1.121 and 1.121, respectively.

Nearly the same results are obtained by using these methods in the case of PLAT362 which is obtained from oceanographic models of Platzman. With

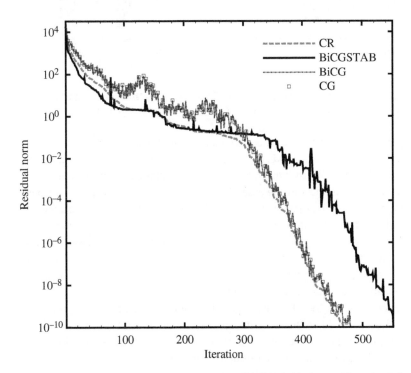

Figure 4.11   The convergence behavior of CR, BiCGSTAB, BiCG, and CG methods in solution of a linear system with symmetric indefinite coefficient matrix BCSSTK22.

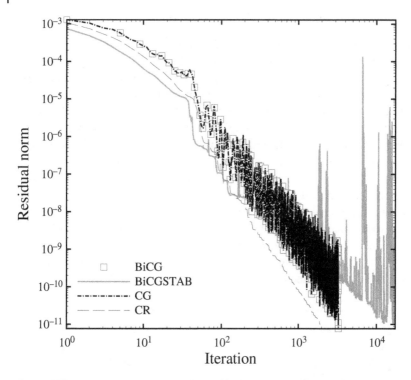

Figure 4.12 The convergence behavior of CR, BiCGSTAB, BiCG, and CG methods in solution of a linear system with symmetric indefinite coefficient matrix PLAT362.

convergence criterion 10^{-10}, for the Euclidean norm of residual vectors, the maximum error value for all methods is 0.1, so by reducing the convergence criterion to 10^{-11}, the maximum error becomes 0.01. The convergence behavior of the CR, BiCGSTAB, BiCG, and CG in the case of PLAT362 are given in Figure 4.12. The CGS method is also diverged in this case. The BiCGSTAB is not recommended because of severe fluctuations in its convergence process. Details of maximum error along with convergence CPU-time of each method in this case is given in Table 4.3. As is evident, among these methods, the CG method has the least convergence CPU-time. Applying the CG method in cases of asymmetric matrices can either lead to divergence or enormous number of iterations for satisfying the convergence criterion; therefore this method is not recommended for asymmetric cases. Definiteness of coefficient matrix is the sufficient condition for existence of the minimum value for its quadratic form, hence it does not necessarily imply inefficiency of the CG method in cases of indefinite matrices.

Table 4.3 Maximum exact error and processing time in the solution of systems with coefficient matrices BCSSTK22, and PLAT362.

	BCSSTK22		PLAT362	
	Maximum Error	Processing Time [sec]	Maximum Error	Processing Time [sec]
CG	2.19e-13	0.01	0.04022	0.42
BiCG	2.19e-13	0.03	0.01022	0.54
BiCGSTAB	6.15e-11	0.04	0.08457	4.50
CR	1.14e-12	0.03	0.08345	0.49

4.2 Preconditioning

The convergence rate of iterative methods depends on spectral characteristics of the coefficients matrix (Chen, 2005). Thus it is possible to transform the equation system into another equivalent system with more appropriate spectral characteristics and the same solution (Saad, 1996). The matrix which provides this transformation is called the preconditioner matrix. For instance, by considering M as an approximation of the coefficient matrix A, the transformed system is as follows

$$M^{-1}Ax = M^{-1}b. \tag{4.1}$$

This system has the same solution with $Ax = b$, but spectral characteristics of the coefficient matrix $M^{-1}A$ may be more desirable (Barrett et al., 1994). There are two approaches for preconditioning. The first approach is devoted to finding matrix M as an approximation to the matrix A such that attaining a solution is easier in the new system. In the second approach we attempt to find matrix M as an approximation to A^{-1} which turns the procedure into a simple product of $M^{-1}b$. Most preconditioners use the first approach.

Given that the use of a preconditioner in a iteration based method, imposes an additional computational cost on each iteration, both in construction and implementation, there is always a conflict between this further computational cost and the convergence speed up. Some preconditioners, such as SSOR, does not impose an additional computational cost for construction while some others such as incomplete factorization requires considerable computational cost.

4.2.1 Preconditioned Conjugate Gradient Method

Consider a symmetric positive-definite matrix A and assume that a preconditioner such as M exists. The preconditioner M is a symmetric positive-definite matrix expressing an approximation of A. From the practical point of view, the matrix M should be derived such that solving the linear system $Mx = b$ does not impose much computational cost, because in each iteration step of the preconditioned methods the mentioned system must be solved. It should be noted that the linear system 4.1 or its equivalent

$$AM^{-1}u = b, \quad x = M^{-1}u \tag{4.2}$$

will not remain symmetric. Here we want to point out some approaches to keep symmetry in the problem. The matrix M can be expressed as an incomplete Cholesky factorization as follows

$$M = LL^T.$$

A simple approach to keep symmetry is to split the preconditioner into two parts, left and right. This means Equation (4.2) can be written as Equation (4.3) which includes a symmetric positive-definite matrix (Saad, 1996; Montagne and Ekambaram, 2004).

$$L^{-1}AL^{-T}u = L^{-1}b, \quad x = L^{-T}u. \tag{4.3}$$

However, splitting the preconditioner in this way is not necessary to maintain symmetry.

The matrix $M^{-1}A$ is called self-adjoint whenever it satisfies the weighted inner product of M, that is

$$(x,y)_M = (Mx,y) = (x,My),$$

which means

$$(M^{-1}Ax,y)_M = (Ax,y) = (x,Ay) = (x,M(M^{-1}A)y) = (x,M^{-1}Ay)_M.$$

Thus, the Euclidean inner product can be replaced by M the weighted inner product in the CG method. If $r^{(i)} = b - Ax^{(i)}$ is the residual vector in the initial linear system and $z^{(i)} = M^{-1}r^{(i)}$ is the residual vector belonging to the preconditioned system, then the arithmetic operations of the preconditioned CG method, except the initial step, are as follow

1. $\alpha_i \leftarrow (z^{(i)}, z^{(i)})_M / (M^{-1}Ap^{(i)}, p^{(i)})_M$
2. $x^{(i+1)} \leftarrow x^{(i)} + \alpha_i p^{(i)}$
3. $r^{(i+1)} \leftarrow r^{(i)} - \alpha_i Ap^{(i)}$
4. $z^{(i+1)} \leftarrow M^{-1}r^{(i+1)}$
5. $\beta_i \leftarrow (z^{(i+1)}, z^{(i+1)})_M / (z^{(i)}, z^{(i)})_M$
6. $p^{(i+1)} \leftarrow z^{(i+1)} + \beta_i p^{(i)}$.

Since $(z^{(i)}, z^{(i)})_M = (r^{(i)}, z^{(i)})$ and $(M^{-1}Ap^{(i)}, p^{(i)})_M = (Ap^{(i)}, p^{(i)})$, weighted inner product of M is not explicitly calculated. Thus the algorithm of the preconditioned CG method can be expressed as Algorithm 4.1.

Algorithm 4.1 Preconditioned Conjugate Gradient Method

```
 1: Initialization:
 2:     r^(0) = b - Ax^(0)
 3:     z^(0) = M^(-1)r^(0)
 4:     p^(0) = z^(0)
 5: while The convergence criteria is not met do
 6:     α_i = (r^(i), z^(i))/(Ap^(i), p^(i))
 7:     x^(i+1) = x^(i) + α_i p^(i)
 8:     r^(i+1) = r^(i) - α_i Ap^(i)
 9:     z^(i+1) = M^(-1)r^(i+1)
10:     β_i = (r^(i+1), z^(i+1))/(r^(i), z^(i))
11:     p^(i+1) = z^(i+1) + β_i p^(i)
12: end while
```

4.2.2 Preconditioning With the ILU(0) Method

Consider a sparse matrix such as A. Upper triangular matrix U and lower triangular matrix L are derived by incomplete LU factorization such that residual matrix $R = LU - A$ meets some particular conditions. One of these conditions is the existence of zero elements in special positions in the matrix R. Here, only the ILU(0) factorization, the simplest form of ILU preconditioning, will be discussed. The ILU factorization without fill-in elements is referred to as ILU(0) with exactly the same zeros distribution as the matrix A. Consider a 32×32 matrix A corresponding to a grid $n_x \times n_y = 8 \times 4$. Figure 4.13 shows the matrix

Figure 4.13 The ILU(0) factorization for a five-point stencil matrix (Saad, 1996).

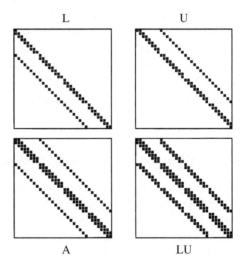

L   U

A   LU

Algorithm 4.2 ILU(0) factorization

```
 1: for i = 2, ..., n do
 2:    for k = 1, ..., i − 1 do
 3:       if aᵢₖ ≠ 0 then aᵢₖ = aᵢₖ/aₖₖ
 4:          for j = k + 1, ..., n do
 5:             if aᵢⱼ ≠ 0 then aᵢⱼ = aᵢⱼ − aᵢₖaₖⱼ
 6:             end if
 7:          end for
 8:       end if
 9:    end for
10: end for
```

L and U that are lower and upper triangular matrices respectively that have the same non-zero entry distribution pattern with corresponding parts in A. The LU multiplication results in a matrix the pattern of which is illustrated in Figure 4.13. The non-zero patterns of A and LU are not exactly the same since LU possess two extra diagonals where their deviations from the main diagonal are $n_x - 1$ and $-n_x + 1$. Elements of these two diagonals are referred to as fill-in elements. However, if these fill-in elements are ignored, the matrices L and U can be found such that the other diagonals of the matrix LU and A are equal. Therefore a general definition for ILU(0) factorization can be expressed. Strictly speaking, when the entries of the matrix $A - LU$ at the position of non-zero elements of A are zero, the lower triangular matrix L and upper triangular matrix U are called ILU(0) factors. These constraints do not provide a unique definition of ILU(0) factors, since a large number of L and U matrices can be found to satisfy these conditions. Factorization algorithm of ILU(0) is given in Algorithm 4.2 (Akhunov et al., 2013). The following subsection is devoted to the numerical analysis of preconditioning with ILU(0) factorization technique in convergence of the Krylov subspace methods.

4.2.3 Numerical Solutions Using Preconditioned Methods

The algorithm of other preconditioned methods can be also overwritten, inspired by the remarks on the preconditioned CG method, which we will avoid here because of the length of the content. By developing codes for the preconditioned solvers algorithm, considerable and significant results were extracted, some of which are mentioned here.

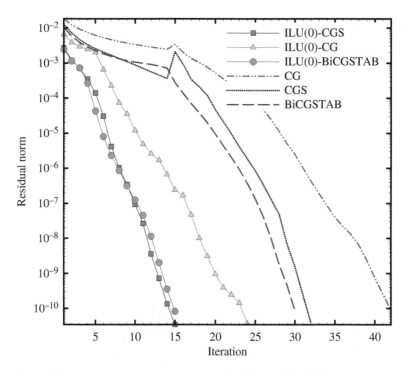

Figure 4.14 The convergence behavior of CG, CGS, and BiCGSTAB methods with and without use of the ILU(0) preconditioner in the case of GR3030.

As stated earlier, preconditioning the coefficient matrix will improve the spectral characteristics of the system. It consequently improves the smoothness of the convergence, iteration numbers, convergence CPU-time, reduction factor, etc. Here we will only study the effect of preconditioning by the factorization method. Figure 4.14 compares the convergence of some preconditioned and unpreconditioned Krylov subspace methods in solving the GR3030 equation system. The results show that the number of iterations required for convergence in solving the GR3030 equations system is reduced by approximately half and the mean reduction factor values will also decrease significantly, see Table 4.4. As mentioned before, the CGS and BiCGSTAB methods will diverge in the case of the NOS7 equation system. Preconditioning the system of equations, or better yet applying the preconditioned solvers ILU(0)-CGS and ILU(0)-BiCGSTAB convergence is realized. Figure 4.15 shows the convergence process of the CG method and preconditioned versions of CG, CGS and BiCGSTAB methods. As can be seen in this figure, convergence of the conventional CG method undergoes severe fluctuations while preconditioning will substantially smooth

Table 4.4 Mean reduction factors of CG, CGS, and BiCGSTAB methods and their preconditioned versions in different systems.

	CG	P-CG	CGS	P-CGS	BiCGSTAB	P-BiCGSTAB
GR3030	0.65	0.45	0.70	0.28	0.57	0.30
NOS7	2.32	1.00	-	8.87	-	1.84
SHERMAN4	0.96	0.78	71.17	0.76	0.88	0.55
ORSREG1	-	0.99	564978.76	40.61	2.99	1.33
BCSSTK22	1.12	0.65	-	237.90	1.11	0.83

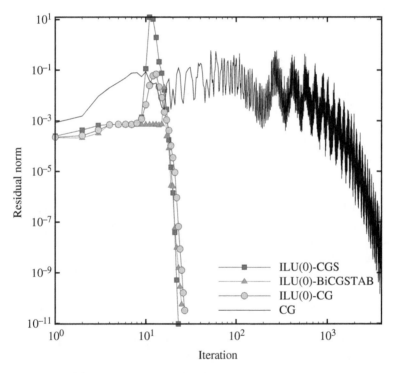

Figure 4.15 The convergence behavior of CG method and preconditioned versions of CG, CGS, and BiCGSTAB methods in the case of NOS7.

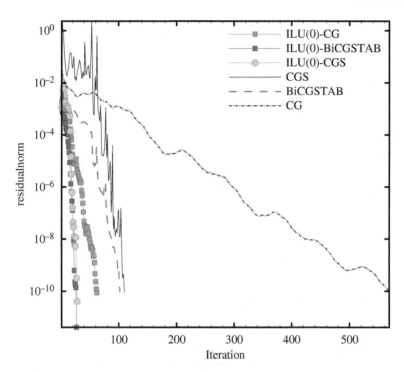

Figure 4.16 The convergence behavior of CG, CGS, and BiCGSTAB methods with and without use of ILU(0) preconditioner in the case of SHERMAN4.

the convergence behavior and reduce the number of required iterations from order of 10^3 to the order of 10. The convergence behavior of the CG, CGS, and BiCGSTAB methods along with their preconditioned versions in solution of a system of equations with asymmetric coefficient matrix SHERMAN4 is shown in Figure 4.16. As mentioned previously, the CG method is not suitable for solving an asymmetric system of equations and if it converges, its convergence rate is slow and the number of required iterations is several times that of other methods. Using preconditioned versions of CG, CGS, and BiCGSTAB solvers, the number of iterations is reduced and the convergence process becomes more regular; one can see the considerable changes of the mean reduction factors in Table 4.4. The effect of preconditioning on the convergence of the methods in solving the asymmetric system of equations with the coefficient matrix ORSREG1 is remarkable as shown

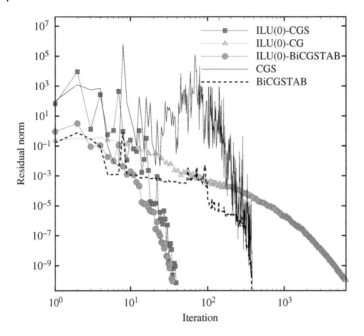

Figure 4.17 The convergence behavior of CG and BiCGSTAB methods in comparison with preconditioned versions of CG, CGS, and BiCGSTAB in the case of ORSREG1.

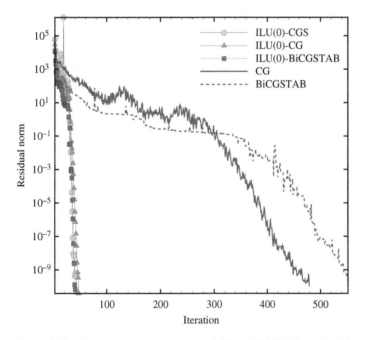

Figure 4.18 The convergence behavior of CG and BiCGSTAB methods in comparison with preconditioned versions of CG, CGS, and BiCGSTAB in the case of BCSSTK22.

Table 4.5 Processing time of CG, CGS, and BiCGSTAB methods and their preconditioned versions in different systems.

	CG	P-CG	CGS	P-CGS	BiCGSTAB	P-BiCGSTAB
GR3030	0.01	0.01	0.01	0.01	0.01	0.01
NOS7	0.9	0	-	0.01	-	0.01
SHERMAN4	0.18	0.01	0.07	0.01	0.06	0.01
ORSREG1	-	4.85	0.49	0.04	0.48	0.06
BCSSTK22	0.01	0	-	0	0.04	0

in Figure 4.17. Table 4.4 indicates that the mean reduction factor of ILU(0)-CGS method in the case of ORSREG1 is not desirable but is remarkably low compared to unpreconditioned CGS method.

Figure 4.18 indicates that in solving the system of equations with symmetric indefinite coefficient matrix BCSSTK22, by preconditioning the coefficients matrix, the CGS method converges and the number of iterations required for the convergence of the CG and BiCGSTAB methods is reduced by approximately one tenth.

There is a comparison between convergence CPU-time of some conventional and preconditioned methods in Table 4.5. Although ILU(0) factorization and its implementation are time consuming, it does play a significant role in reducing convergence CPU-time or essentially converging the solvers.

4.3   Numerical Solution of Systems Using GMRES*

This section is devoted to the numerical solution of the equations system NOS7 and BCSSTK22 using the GMRES* methods, where the inner iteration limit is set to 20. In the GMRES* method, other Krylov subspace solvers act as preconditioner that, after a certain number of iterations, their residual vectors in an orthogonalization process form the orthogonal basis of the Krylov subspace (Van der Vorst, 2003). As stated in Section 4.1.1, BiCGSTAB method diverges in the case of the NOS7 system and CGS method is not recommended in the cases of the NOS7 and the BCSSTK22 systems. However, if these two solvers are used as preconditioners in the GMRES* algorithm, one can obtain remarkably improved

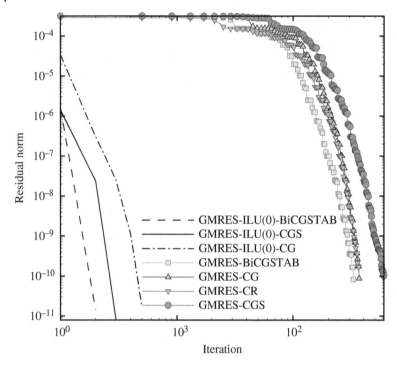

Figure 4.19   The convergence behavior of the GMRES∗ solvers in the case of NOS7.

results in solving these two system of equations. Figure 4.19 and Figure 4.20 depict the convergence trend of the GMRES∗ method in solution of the NOS7 and the BCSSTK22 systems. Maximum error and CPU-time of each method is given in Table 4.6.

4.4   Storage Formats and CPU-Time

Choosing an appropriate storage format for sparse matrices has a great impact on the CPU-time in solving the equation system using the Krylov subspace method. In Section 2.4.2 some common storage formats of sparse matrices are introduced. In this section, four formats CSR, MSR, diagonal, and Ellpack-Itpack are compared in terms of CPU-time. We need a benchmark to compare storage formats. In many computational fluid dynamics problems, we deal with equation systems with diagonal matrices that can be derived from an elliptic equation such as Poisson's equation.

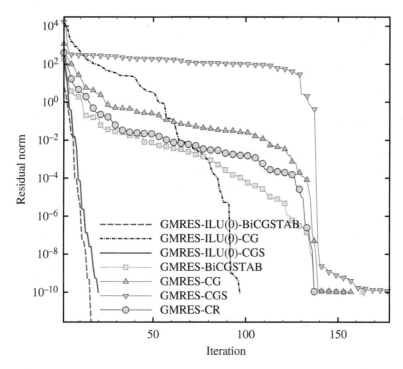

Figure 4.20 The convergence behavior of the GMRES∗ solvers in the case of BCSSTK22.

Table 4.6 Maximum exact error and processing time in the solution of systems with coefficient matrices NOS7 and BCSSTK22 using solvers GMRES∗.

	NOS7		BCSSTK22	
	Processing Time [sec]	Maximum Error	Processing Time [sec]	Maximum Error
GMRES-CG	2.44	1.70e-06	0.06	6.60e-10
GMRES-CGS	6.19	2.04e-04	0.09	1.66e-09
GMRES-BiCGSTAB	2.09	8.38e-07	0.07	5.01e-09
GMRES-CR	2.65	8.89e-07	0.06	3.99e-09
GMRES-ILU(0)-CG	0.01	7.35e-10	0.04	4.47e-11
GMRES-ILU(0)-CGS	0.01	2.96e-10	0.01	4.95e-14
GMRES-ILU(0)-BiCGSTAB	0.01	2.00e-09	0.01	1.71e-13

Consider a Poisson's equation, Equation (4.4), on the rectangular domain $D = [0, 2\pi] \times [0, 2\pi]$.

$$u_{xx} + u_{yy} = f(x, y)|_D,$$
$$u(x, y)|_{\partial D} = \cos(x + y). \tag{4.4}$$

The source function of this equation is selected in a way that the exact solution is equal to $\cos(x + y)$. Using the second order central difference approximation on different structured Cartesian grids, results in the linear systems with large sparse diagonal matrices. A zero vector is set as the initial guess. The convergence criteria is considered to be met when the Euclidean norm of the residuals vector is less than 10^{-10}.

Matrix–vector multiplication is a fundamental and time consuming part of arithmetic operations which has also a significant role in all Krylov subspace methods. On the other hand, discretizations mostly lead to large sparse matrices. Storing such a huge number of zero elements along with other non-zero elements is not a rational approach (Press et al., 1992). Hence, implementation of different

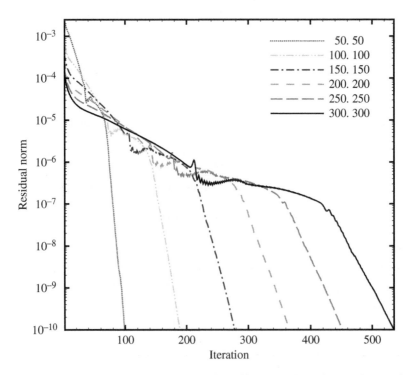

Figure 4.21 The convergence behavior of the CG method for various grid resolutions.

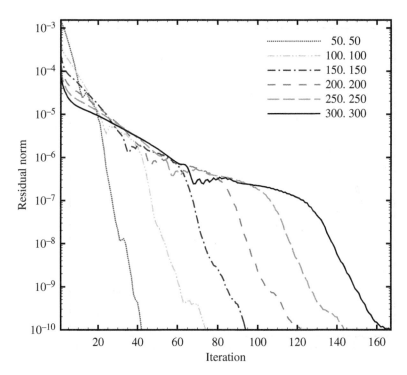

Figure 4.22 The convergence behavior of the ILU(0)-CG method for various grid resolutions.

storage formats and using them in the arithmetic operations plays a remarkable role in reducing the CPU-time.

The CG method contains one matrix–vector multiplication in each iteration. This method can be a suitable test case for studying and comparing the convergence CPU-time when the different algorithms of matrix–vector multiplication are applied. Figures 4.21 and 4.22 show the convergence behavior of the CG method and its preconditioned version ILU(0)-CG. These two methods are applied for solving Equation (4.4) on a computational domain with various grid resolutions. From the numerical investigations it can be seen that in the CG method, the number of required iterations for convergence is proportional to $(\Delta h)^{-1}$ (Axelsson and Barker, 1984), where Δh is the spatial step. Figures 4.21 and 4.22 show that implementation of ILU(0) preconditioner reduces the number of iterations to approximately half and even up to one third for the grid resolutions of 50×50 and 300×300, respectively.

The required memory for storing a five diagonal sparse matrix, resulting from discretization of a Poisson's equation, with four formats, CSR, MSR, diagonal, and Ellpack-Itpack is plotted versus rank of matrix in Figure 4.23. It is evident

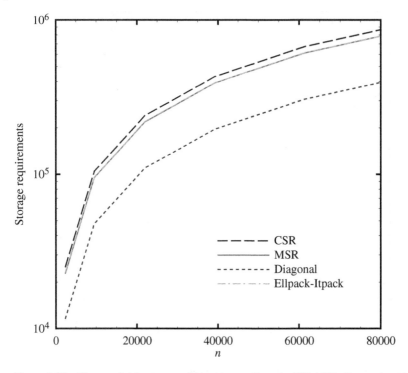

Figure 4.23 The required memory of four storage formats CSR, MSR, diagonal, and Ellpack-Itpack, versus the rank of matrix.

from Figure 4.23 that the diagonal format requires the least storage memory in the case of five diagonal sparse matrix. The algorithms of matrix–vector multiplication are used along with different storage formats in the developed codes. This allows comparison of the storage format's effects on convergence CPU-time. According to the results in Figures 4.24 and 4.25, the diagonal format is the most appropriate choice for storing the diagonal matrices and performing the arithmetic operations. The CSR format has the lowest efficiency and is not recommended for use in matrix–vector multiplication, but because of its simplicity it is usually used as an input and intermediate format in developed codes. It should be emphasized that the latter remarks are only valid for the special case of the current problem and diagonal matrices. Therefore, choosing a suitable format is completely dependent on the matrix sparsity pattern, and a general judgment cannot be made for other matrices.

In this book, various matrix–vector multiplication algorithms, as the kernel of the Krylov subspace methods, are applied in developed solvers structure using the compressed formats. Along with development of more efficient storage formats,

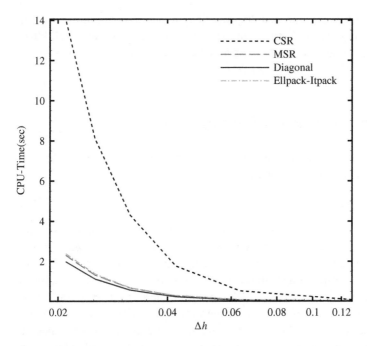

Figure 4.24 The CPU-time of the CG method versus grid size Δh for four storage formats CSR, MSR, diagonal, and Ellpack-Itpack.

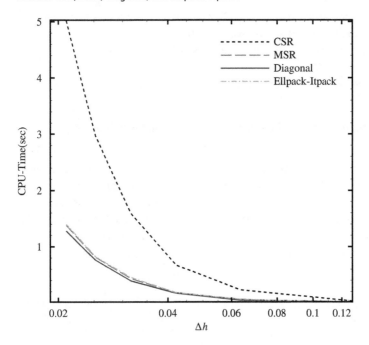

Figure 4.25 The CPU-time of the ILU(0)-CG method versus grid size Δh for four storage formats CSR, MSR, diagonal, and Ellpack-Itpack.

there are comprehensive studies in improvement of matrix–vector multiplication algorithms in computational science literature (Shahnaz and Usman, 2007; Pinar and Heath, 1999; Toledo, 1997; Yzelman and Bisseling, 2009; Im and Yelick, 2001; Haase et al., 2007; Vuduc and Moon, 2005).

4.5 Solution of Singular Systems

Consider a partial differential equation on a domain Γ with pure Neumann boundary conditions along the entire boundaries $\delta\Gamma$, i.e.,

$$u_{xx} + u_{yy} = f(x,y)|_{\Gamma}, \quad \frac{\partial u}{\partial n}|_{\delta\Gamma} = 0. \tag{4.5}$$

Discretization of Equation (4.5) leads to a singular consistent linear system. Because of the singularity of the matrix A in the linear system $Ax = b$, the solution of this system is determined up to a constant (Van der Vorst, 2003). Here the eigenvector corresponding to the eigenvalue of zero is a vector whose elements are all one. If the linear system $Ax = b$ is consistent, the vector b is perpendicular to the eigenvector corresponding to the zero eigenvalue (Van der Vorst, 2003) and can be simply solved using the Krylov subspace methods. A linear system is said to be consistent if there exists at least one solution for it. All of the following statements meet the definition of consistency

- If $[A|b]$ is an augmented matrix, then by Gaussian elimination of rows, a row will never be such as $(0\ 0\ 0\ \cdots\ 0\ |\ \alpha)$, $\quad \alpha \neq 0$.
- b is a non-basic column in the augmented matrix $[A|b]$.
- $rank\ ([A|b]) = rank\ (A)$.
- b is a combination of the basic columns in the matrix A.

4.5.1 Solution of Poisson's Equation with Pure Neumann Boundary Conditions

In this subsection, a singular linear system is derived by discretization of a Poisson's equation with pure Neumann boundary conditions. By selecting 250×250, 500×500, and 1000×1000 computational grids, we evaluate different Krylov subspace methods and investigate the effect of parameters such as grid resolution and convergence criteria on CPU-time and maximum error.

Consider a Poisson's equation

$$u_{xx} + u_{yy} = f(x,y)|_{D},$$
$$\frac{\partial u(x,y)}{\partial n}|_{\partial D} = 0, \tag{4.6}$$

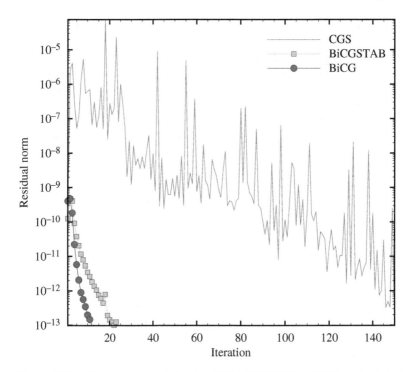

Figure 4.26 The convergence behavior of CGS, BiCGSTAB, and BiCG methods in the solution of the Poisson's equation with the pure Neumann boundary conditions on a 250×250 grid.

on a square domain of $D = [0, 2\pi] \times [0, 2\pi]$. We choose the source function $f(x, y)$ such that the exact solution equals $(0.5 - \cos(x))(0.5 - \cos(y))$. Equation (4.6) is discretized on a 250×250 grid using a second order central difference approximation which results in a system of equations with an asymmetric sparse matrix. The convergence behavior of the CGS, BiCGSTAB, and BiCG methods in solving this system is shown in Figure 4.26. In this problem, the convergence criterion is the reduction of residuals vector Euclidean norm below 10^{-13}. The CG method is not able to reduce the residuals vector norm below this value and the CR method cannot meet the criterion in the limited number of iterations and also its convergence CPU-time significantly increases (see the last line of Table 4.7). Figure 4.26 shows that BiCGSTAB method has smoother convergence compared to the CGS method. Although the number of required iterations for convergence of the BiCGSTAB method is approximately twice that of the BiCG method, the CPU-time is half that (see lines 8 and 16 in Table 4.7). Although the CG method cannot reduce the residual vector norm to less than 10^{-11}, its convergence time is favorable.

Table 4.7 Maximum exact error and processing time for various convergence criteria of solvers in the solution of Equation (4.6) discretized on a 250 × 250 grid.

	Convergence Criteria (Residuals Vector Euclidean Norm)	Maximum Error	Processing Time [sec]
CG	10e-10	0.0028936	0.03
	10e-11	0.0028936	0.03
	10e-12	-	-
	10e-13	-	-
BiCG	10e-10	0.030256	0.10
	10e-11	0.0079307	0.10
	10e-12	0.0012313	0.14
	10e-13	0.0001833	0.24
CGS	10e-10	0.00011634	0.43
	10e-11	0.000108043	0.46
	10e-12	0.000106284	0.68
	10e-13	0.00010609	0.68
BiCGSTAB	10e-10	0.00910682	0.04
	10e-11	0.00033356	0.06
	10e-12	0.00011953	0.07
	10e-13	0.00010776	0.12
CR	10e-10	0.0027912	0.03
	10e-11	0.0027912	0.03
	10e-12	0.0007546	65.59
	10e-13	0.0001319	254.35

According to the numerical results of Table 4.7, it can be concluded that the BiCGSTAB and the CGS methods generally have the least CPU-time and lowest maximum exact error, respectively. Since it is not possible to evaluate and compare the performance of all developed solvers for different convergence criteria, Figures 4.27 and 4.28 are devoted to comparing the convergence process of a limited number of solvers for a particular convergence criterion, and we provide more detailed information in Tables 4.8 to 4.11. Figure 4.27 shows the convergence process of CG, BiCG, CGS, BiCGSTAB, and CR methods in solving the equation system obtained by discretization of Equation (4.6) on a 500 × 500 grid. Here, the convergence criterion is to reduce the Euclidean norm of the residuals vector to 10^{-14}. To solve this problem, the BiCGSTAB method with

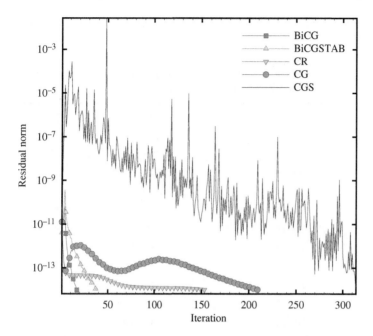

Figure 4.27 The convergence behavior of CG, BiCG, CGS, BiCGSTAB, and CR methods in the solution of the Poisson's equation with the pure Neumann boundary conditions on a 500×500 grid.

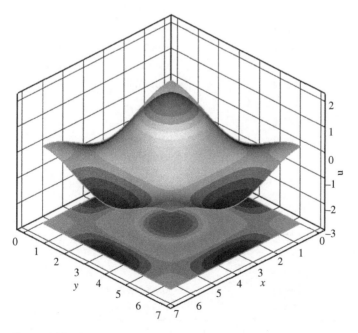

Figure 4.28 The distribution of the function *u* obtained from solving the Equation (4.6).

Table 4.8 Maximum exact error and processing time for various convergence critera of solvers in the solution of Equation (4.6) discretized on a 500 × 500 grid.

	Convergence Criteria (Residuals Vector Euclidean Norm)	Maximum Error	Processing Time [sec]
CG	10e-11	0.001829	0.14
	10e-12	0.001829	0.14
	10e-13	0.001829	0.18
	10e-14	0.000020917	2.65
BiCG	10e-11	0.16303	0.45
	10e-12	0.02012	0.62
	10e-13	0.003465	0.98
	10e-14	0.000417792	1.48
CGS	10e-11	0.21	0.14
	10e-12	0.000027	5.97
	10e-13	0.000026558	5.91
	10e-14	0.000026384	6.89
BiCGSTAB	10e-11	0.16	0.14
	10e-12	0.00056299	0.39
	10e-13	0.00009554	0.57
	10e-14	0.000029344	0.9
CR	10e-11	0.002813	0.2
	10e-12	0.002813	0.2
	10e-13	0.00180416	0.21
	10e-14	0.000371574	3.43
ILU(0)-CG	10e-11	0.01499	2.26
	10e-12	0.00201599	3.35
	10e-13	0.00044344	4.58
	10e-14	0.0000321362	6.78
ILU(0)-CGS	10e-11	0.005386	2.9
	10e-12	0.000476	3.72
	10e-13	0.00004762	4.63
	10e-14	0.000046596	4.77
ILU(0)-BiCGSTAB	10e-11	0.2552	0.6396
	10e-12	0.00459	1.17
	10e-13	0.00058826	1.68
	10e-14	0.000202906	2.46

Table 4.9 Maximum exact error and processing time for various convergence criteria of GMRES* solvers in the solution of Equation (4.6) discretized on a 500 × 500 grid.

	Convergence Criteria (Residuals Vector Euclidean Norm)	Maximum Error	Processing Time [sec]
GMRES-CG	10e-11	0.0005919	1.1
	10e-12	0.0005919	1.1
	10e-13	0.0005920	1.13
	10e-14	0.0001827	7.64
GMRES-CGS	10e-11	0.000097	2.94
	10e-12	0.000097	2.94
	10e-13	0.0000347	26.20
	10e-14	0.00551064	7.14
	(20 inner iterations)		
	10e-14	0.187248	11.68
GMRES-BiCGSTAB	10e-11	0.000027	1.56
	10e-12	0.0000272	1.52
	10e-13	0.0000272	1.52
	10e-14	0.000028979	1.17
GMRES-CR	10e-11	0.000494	1.59
	10e-12	0.000494	1.57
	10e-13	0.000494	1.59
	10e-14	0.000029185	11.38
GMRES-ILU(0)-CG	10e-11	0.007375	8.81
	10e-12	0.0033782	11.74
	10e-13	0.0000837	16.27
	10e-14	0.0115685	12.18
GMRES-ILU(0)-CGS	10e-11	0.00632	7.03
	10e-12	0.00631	7.16
	10e-13	0.0001206	13.97
	10e-14	0.000090634	19.89
GMRES-ILU(0)-BiCGSTAB	10e-11	0.000262	2.57
	10e-12	0.000261	2.48
	10e-13	0.0002618	2.48
	10e-14	0.00029928	5.61

Table 4.10 Maximum exact error and processing time for various convergence critera of solvers in the solution of Equation (4.6) discretized on a 1000×1000 grid.

	Convergence Criteria (Residuals Vector Euclidean Norm)	Maximum Error	Processing Time [sec]
CG	10e-12	0.33391	0.48
	10e-13	0.0029066	0.59
	10e-14	0.0029066	0.68
	10e-15	0.00006971	16.55
BiCG	10e-12	0.33391	0.39
	10e-13	0.078916	3.54
	10e-14	0.010231	5.61
	10e-15	0.0007705	8.75
CGS	10e-12	0.20905	0.57
	10e-13	0.0000094301	65.30
	10e-14	0.0000066487	81.52
	10e-15	0.0000064654	83.53
BiCGSTAB	10e-12	0.1854	0.56
	10e-13	0.0015031	2.05
	10e-14	0.0002479	3.35
	10e-15	0.00004229	5.52
CR	10e-12	0.41246	0.57
	10e-13	0.0005323	0.74
	10e-14	0.0027463	0.92
	10e-15	0.00062635	9.48
ILU(0)-CG	10e-12	0.0646519	14.85
	10e-13	0.00509486	26.03
	10e-14	0.00069373	38.09
	10e-15	0.000043655	52.93
ILU(0)-CGS	10e-12	0.0062863	20.10
	10e-13	0.00051138	26.87
	10e-14	0.0000324185	35.33
	10e-15	0.0000230664	35.55
ILU(0)-BiCGSTAB	10e-12	0.36148	0.9672
	10e-13	0.036234	5.08
	10e-14	0.0034451	7.37
	10e-15	0.000372704	10.71

Table 4.11 Maximum exact error and processing time for various convergence critera of GMRES∗ solvers in the solution of Equation (4.6) discretized on a 1000 × 1000 grid.

	Convergence Criteria (Residuals Vector Euclidean Norm)	Maximum Error	Processing Time [sec]
GMRES-CG	10e-12	0.0029068	2.35
	10e-13	0.0029068	2.37
	10e-14	0.0018567	20.63
	10e-15	0.0018567	20.63
GMRES-CGS	10e-12	0.0116060	11.32
	10e-13	0.010572	28.71
	10e-14	0.010572	28.67
	10e-15	0.010572	28.71
GMRES-BiCGSTAB	10e-12	0.00343874	3.04
	10e-13	0.00343874	3.01
	10e-14	0.0105115	27.36
	10e-15	0.0105115	27.34
GMRES-CR	10e-12	0.0028863	3.18
	10e-13	0.0028863	3.18
	10e-14	0.0028863	3.15
	10e-15	0.00019403	5.85
GMRES-ILU(0)-CG	10e-12	0.3411	9.20
	10e-13	0.2516	30.70
	10e-14	0.2516	30.73
	10e-15	0.2516	30.66
GMRES-ILU(0)-CGS	10e-12	1.2719	44.95
	10e-13	1.2719	44.95
	10e-14	1.2719	44.69
	10e-15	1.2719	44.67
GMRES-ILU(0)-BiCGSTAB	10e-12	0.7984	13.21
	10e-13	0.055886	26.23
	10e-14	0.012691	44.07
	10e-15	0.012691	44.02
	10e-12	0.6257	15.42
GMRES-ILU(0)-CGS	10e-13	0.00067551	76.70
(40 inner iteration)	10e-14	0.00067551	76.47
	10e-15	0.00067551	76.51
	10e-12	0.0100036	7.84
GMRES-ILU(0)-BiCGSTAB	10e-13	0.0100036	7.84
(40 inner iteration)	10e-14	0.0033404	15.21
	10e-15	0.0017345	75.42

39 iterations and CPU-time of 0.9 seconds is desirable. The accuracy required in solving a system can play a decisive role in selecting a suitable solver. For example, to obtain an accuracy of order 10^{-3}, the CG solver with convergence criterion 10^{-11} suffices which has the lowest processing time among the solvers presented in Table 4.8. To achieve an accuracy of 10^{-5}, the BiCGSTAB is proposed with convergence criterion of 10^{-13} and processing time of 0.57 seconds.

Table 4.9 presents the maximum exact error and the processing time in solving the system of equations using the GMRES* method. In all GMRES* solvers presented in Table 4.9, the number of inner iterations is 40. The processing time and the maximum exact error will be considerably changed by altering the number of inner iterations. For example in the case of the GMRES-CGS solver, reducing the number of inner iterations from 40 to 20 will have a positive effect on the maximum exact error; see lines 8 and 9 in Table 4.9. In some cases, decrease in the maximum exact error and the processing time will be achieved by increase in the number of inner iterations; for example see lines 33, 34 and 35 of Table 4.11. Thus it is not possible to explain the effect of the inner iteration number on the maximum exact error or the processing time. Figure 4.28 shows the distribution of the multivariable function u resulting from the solution of Equation (4.6) using the Krylov subspace methods. Figures 4.29 to 4.31 shows the error distribution obtained from the solution of a system of equations arising from Equation (4.6)

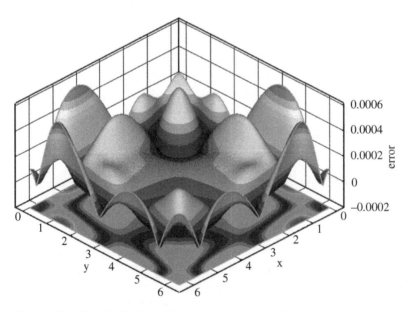

Figure 4.29 The distribution of the exact error obtained from solving the Equation (4.6) using the ILU(0)-BiCGSTAB solver.

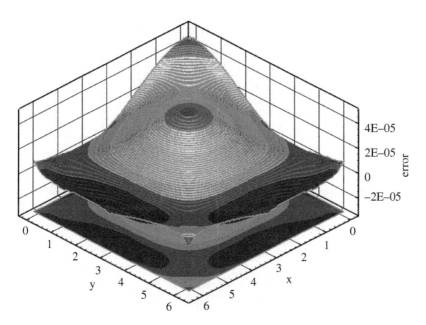

Figure 4.30 The distribution of the exact error obtained from solving the Equation (4.6) using the ILU(0)-CGS solver.

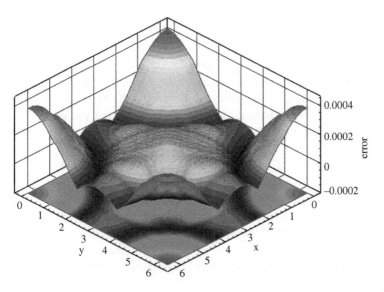

Figure 4.31 The distribution of the exact error obtained from solving the Equation (4.6) using the ILU(0)-CG solver.

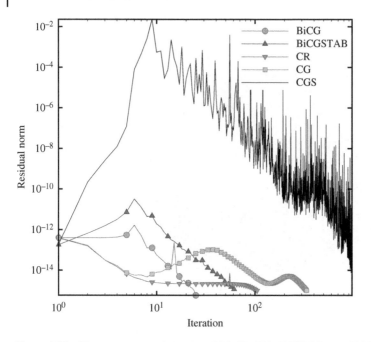

Figure 4.32 The convergence behavior of CG, CR, CGS, BiCGSTAB, and BiCG methods in the solution of the Poisson's equation with pure Neumann boundary conditions on a 1000 × 1000 grid.

when the ILU(0)-BiCGSTAB, ILU(0)-CGS, and ILU(0)-CG methods are applied. It should be noted that the discretization is performed on a 500 × 500 grid and the convergence criterion is 10^{-13}. Methods of error reduction and the unique nature of convergence in different Krylov subspace methods make the different error distributions. Figure 4.32 shows the comparison of the convergence process of the CG, BiCG, CGS, BiCGSTAB, and CR methods over a 1000 × 1000 grid which yields results similar to Figure 4.27. The CG method with the convergence criterion of 10^{-13} is able to reduce the exact error with a processing time of 0.59 seconds to the order of 10^{-3} which is highly desirable; see Table 4.10. If we want to achieve an accuracy of 10^{-4} with the lowest processing time, the CR method with the convergence criterion of 10^{-13} can be used. In this case one can reach an exact error of about 0.000 532 3 with a CPU-time of 0.74 seconds. To achieve an accuracy of order 10^{-5}, the BiCGSTAB method with convergence criterion of 10^{-15} is recommended. By using the CGS method, one can achieve the accuracy of 10^{-6} but the processing time is not optimal.

Table 4.11 indicates processing time and maximum exact error when GMRES∗ solvers with 20 inner iterations converge. As mentioned previously, inner iterations number of GMRES∗ has remarkable effect on processing time and maximum exact error. Therefore, the optimal selection of the inner iterations number can be considered as an independent field of study.

4.5.2 Comparison of the Krylov Subspace Methods with the Point Successive Over-Relaxation (PSOR) Method

In this subsection, performance of the Krylov subspace methods are compared with the point successive over-relaxation (PSOR) method. Information on the maximum exact error and CPU-time in the numerical solution of the Poisson's equation with pure Neumann boundary conditions on a 500×500 grid is presented in Table 4.12.

Tables 4.8 and 4.9 show that the GMRES-CGS solver with 40 inner iterations has the maximum processing time and with convergence criterion of 10^{-13} can reduce the maximum exact error to the order of 10^{-5}. It should be noted that processing time can be improved by altering the number of inner iterations. Therefore, due to the accuracy of this method, the processing time can be ignored.

By carefully comparing the results presented in Table 4.12 and comparing it with the information presented in Tables 4.8 and 4.9, it can be concluded that the Krylov subspace solvers have a significantly higher preference over the PSOR method.

Table 4.12 Maximum exact error and processing time for various convergence critera of PSOR solver in the solution of Equation (4.6) discretized on a 500×500 grid.

Convergence Criteria (Euclidean Norm of Iteration Errors Vector)	Maximum Error	Processing Time [sec]
10e-4	0.066 287 5	17.37
10e-5	0.023 644 4	24.57
10e-6	0.013 851 3	32.51
10e-7	0.020 908 4	47.51
10e-8	0.031 836 2	75.70

Exercises

4.1 Consider the Poisson's equation $u_{xx} + u_{yy} = f(x, y)$ on a square domain $[0, 2\pi] \times [0, 2\pi]$ with the exact solution $\cos(x + y)$ which results in the source function $-2\cos(x + y)$.

a) Discretize the equation on a 1000×1000 square grid using a second- and fourth-order central difference approximation which results in the five- and nine-point stencil matrices, respectively. Store the matrices in the coordinate format using your developed FORTRAN or C program in response of the part (a) of Exercise 1 in Chapter 2.

b) Convert the resulting coordinate format to CSR format. Try to apply the subroutine of Appendix A.1.1.

c) Develop a FORTRAN or C program and apply the subroutines of Appendix B to solve the resulting systems of equations in part (a).

4.2 Consider the resulting systems of part (a) of Exercise 1 and apply the subroutines of the Appendix B and Appendix C, and compare the convergence rate and processing time of the CG, CGS, and BiCGSTAB with their ILU(0)-preconditioned versions.

4.3 Consider the resulting systems of part (a) of Exercise 1 and apply the different matrix–vector multiplication algorithm and evaluate the processing time of the GMRES∗ method when the inner iterations of Appendix D are applied as its preconditioners.

4.4 Consider the resulting systems of part (a) of Exercise 1 and apply the stationary solvers point-SOR, line-SOR, point-Gauss-Seidel, and line-Gauss-Seidel as the inner iterations of the GMRES∗ method and compare the rate of convergence and processing time with the conditions that the Krylov subspace methods are used as the inner iterations.

4.5 Consider the resulting systems of part (a) of Exercise 1 and investigate the effect of the inner iterations number on the convergence rate and exterior iterations number of the GMRES∗ method.

4.6 Explain why the CR method has a smoother convergence process than the other methods.

4.7 Consider the Equation (4.6) and discretize it on the square domain $[0, 2\pi] \times [0, 2\pi]$ with grid resolution 1000×1000. Apply the five-point and nine-point stencils and develop a FORTRAN or C program for ILU(0)-preconditioned version of BiCG and try to solve the resulting system of equations when the pure Neumann boundary conditions are set.

a) Compare the convergence trend and processing time of the BiCG with ILU(0)-preconditioned BiCG.

b) Try to use the inner iterations of BiCG and ILU(0)-preconditioned BiCG in GMRES* algorithm and develop a FORTRAN or C program for it.

c) Design an algorithm for the transpose matrix–vector multiplication for the diagonal and Ellpack-Itpack formats and develop a FORTRAN or C program to implement it. Apply the subroutines of Appendix A.2, Appendix A.3 and your developed subroutines in the subroutine of Appendix B.2 and compare the processing time of the BiCG when the various algorithms of the matrix–vector multiplication and their transpose pairs are implemented.

5

Solution of Incompressible Navier–Stokes Equations

5.1 Introduction

The major challenge in solving the governing equations of the incompressible fluid flow is the lack of independent equation for the pressure field. Thus the main idea of different approaches in solution of incompressible Navier–Stokes equations is to deal with this challenge (Ferziger, 1987).

The main idea of the projection methods in the late 1960's was first proposed by Chorin (1968) and independently by Temam (1969), on a more robust mathematical basis, as an appropriate and optimal method of solution to the incompressible Navier–Stokes equations. The use of this set of methods for numerical simulation of incompressible viscous flows with medium and high Reynolds numbers is very common (Rempfer, 2006).

One of the prominent features of the projection methods is their generality and robust mathematical basis compared to other methods which are based on specific solutions or physical phenomena. The projection methods, as the most widely used incompressible flow solvers, are based on the primary variables formulation of the Navier–Stokes equations (McDonough, 2007).

In a general sense, the philosophy of the projection method is based on the assumption that, in an incompressible flow, pressure has no thermodynamical meaning and acts only as a Lagrangian multiplier to hold the assumption of incompressibility of flow (Pozrikidis, 2003). This concept results in a fractional time step discretization scheme in which the velocity and pressure field computations are performed separately. This scheme is a prominent feature of the projection methods.

In the first step, an auxiliary velocity field is computed using the momentum equations regardless of the pressure gradient terms. In the second step, the auxiliary velocity field is mapped into a divergence-free vector space in order to compute the physical velocity and pressure field for the next time step. This method of

Krylov Subspace Methods with Application in Incompressible Fluid Flow Solvers, First Edition.
Iman Farahbakhsh.
© 2020 John Wiley & Sons Ltd. Published 2020 by John Wiley & Sons Ltd.
Companion Website: www.wiley.com/go/Farahbakhs/KrylovSubspaceMethods

solving the governing equations is much more efficient than solving the system of incompressible Navier–Stokes equations which simultaneously contain both pressure and velocity terms. The major difficulty in this method, as detailed below, is to introduce a suitable numerical boundary layer on the auxiliary velocity and the pseudo-pressure fields. The main difficulty in designing and applying the optimum projection methods is the proper reconstruction of the boundary conditions. The projection method which is also known as the fractional step method or the fractional step projection method, provides more efficient solutions for large eddy simulation (LES) with medium and high Reynolds numbers using the operator splitting approach (Rempfer, 2006; Piomelli and Balaras, 2002).

In recent years, there has been increasing attention on the projection methods for direct numerical simulation (DNS) of incompressible viscous flows. It should also be noted that in this approach, different spatial discretization methods may be used such as finite difference, upwind difference scheme, or spectral element methods (McDonough, 2007).

5.2 Theory of the Chorin's Projection Method

All of the projection methods are theoretically based on a theorem which is known as Hodge vector space decomposition or Hodge decomposition and states that any vector function $v \in L^2(\Omega), \Omega \subset \mathbb{R}^d, d = 2, 3$ can be decomposed as (McDonough, 2007)

$$v = \nabla q + \nabla \times \phi \tag{5.1}$$

where q and ϕ are properly smooth. This theorem simply states that any L^2 vector field can be decomposed into two curl-free and divergence-free terms. A special case of this theorem, called Helmholtz vector space mapping, is well known in computational fluid dynamics applications. By imposing smoothness constraint in Helmholtz mapping, a new form of the L^2 vector space results from the Hodge theorem. In Helmholtz–Leray mapping, a special vector space v is considered on the domain $\Omega \subset \mathbb{R}^d, d = 2, 3$ with a boundary condition on $\partial\Omega$; by this assumption, we seek a new definition for vector space v as

$$v = \nabla\phi + u, \quad \nabla.u = 0. \tag{5.2}$$

By taking the divergence of both sides of Equation (5.1) we have

$$\nabla^2\phi = \nabla.v. \tag{5.3}$$

Hence, an elliptic equation is derived for computation of the scalar field ϕ. Equation (5.3) can be directly solved by having a boundary condition ϕ based on the vector space v. If v satisfies no-penetration and no-flux conditions, then by imposing $u.n = 0$ over the boundary $\partial\Omega$, a suitable boundary condition is derived. The inner production of the outward unit normal vector n and Equation (5.2), considering the mentioned condition for u, results in the following boundary condition

$$\nabla\phi.n = v.n \rightarrow \frac{\partial\phi}{\partial n} = v.n \quad \text{on} \quad \partial\Omega \tag{5.4}$$

Equation (5.4) states the Neumann boundary condition for the Poisson's equation ϕ. To ensure an answer to Equation (5.3), the right hand side of the boundary condition, Equation (5.4), must have the following compatibility condition, which will automatically establish the no-penetration and no-flux conditions for the vector v over the boundary $\partial\Omega$.

$$\int v.ndA = 0 \quad \text{on} \quad \partial\Omega. \tag{5.5}$$

5.3 Analysis of Projection Method

The current section is devoted to a detailed analysis of the projection method (McDonough, 2007) which was initially presented by Kim and Moin (1985). Incompressible Navier–Stokes equations in vector notation can be presented as

$$\begin{aligned} u_t + u\nabla.u &= -\frac{1}{\rho}\nabla p + v\Delta u + F \\ \nabla.u &= 0 \end{aligned} \tag{5.6}$$

where v is the kinematic viscosity, u is the velocity vector, p is the pressure and F is the acceleration vector due to field forces such as gravity, electromagnetism, etc. In the incompressible flow momentum equations, the forces induced by non-linear acceleration, pressure, viscous, and body effect terms can change the velocity field. In the fractional step method, which is extensively studied by Janenko (1971), these effects, under the some special conditions, can be independently assumed by considering the small changes of time, Δt.

The first step in most of the basic projection methods is to solve the momentum equations by ignoring the pressure gradient term. Thus Equation (5.6) can be decomposed into two equations

$$\hat{u}_t = -\nabla.(\hat{u}^2) + v\Delta\hat{u} + F, \tag{5.7}$$

$$u_t = -\frac{1}{\rho}\nabla P. \tag{5.8}$$

By this assumption, the pressure gradient form cannot be assumed to be the same as the physical pressure gradient; hence in new relationships, p is substituted by P. Equation (5.7) shows the vector form of the Burgers' equations (McDonough,

2007). Solving this set of equations, in a time step, results in the auxiliary velocity $\hat{\boldsymbol{u}}$. Since the pressure term is ignored in momentum equations, the computed velocity field from Burgers' equations can no longer satisfy the boundary conditions and hence it is distinguished from its physical value by the hat superscript. Equation (5.8) is used to satisfy continuity condition. The basic idea in the projection method is that the pseudo-pressure field $\mathcal{P}$ should be computed in a way that the final velocity field has to be solenoidal or in the other words, it should satisfy the continuity condition. In order to derive pseudo-pressure Poisson's equation, one can take the divergence of both sides of Equation (5.8) which leads to

$$\nabla.\boldsymbol{u}_t = (\nabla.\boldsymbol{u})_t = -\nabla.\frac{1}{\rho}\nabla\mathcal{P},$$

$$\nabla^2\mathcal{P} = -\rho(\nabla.\boldsymbol{u})_t. \tag{5.9}$$

The right hand side term is approximated as

$$(\nabla.\boldsymbol{u})_t \simeq \frac{(\nabla.\boldsymbol{u})^{n+1} - (\nabla.\hat{\boldsymbol{u}})^n}{\Delta t}. \tag{5.10}$$

To derive the Poisson's equation we assume $(\nabla.\boldsymbol{u})^{n+1} = 0$; then Equation (5.9) can be written as

$$\nabla^2\mathcal{P} = \rho\frac{\nabla.\hat{\boldsymbol{u}}}{\Delta t}. \tag{5.11}$$

It is clear that if the pseudo-pressure field is established, then the physical velocity field can be constituted based on the auxiliary velocity field $\hat{\boldsymbol{u}}$ as follows (McDonough, 2007)

$$\boldsymbol{u}^{n+1} = \hat{\boldsymbol{u}} - \frac{\Delta t}{\rho}\nabla\mathcal{P}. \tag{5.12}$$

The fractional step projection method is independent of the integration process for the temporal progression of the momentum equations. Therefore, the final velocity field cannot have the temporal accuracy higher than the first-order. To achieve this accuracy the following method is proposed. For ease of use, forward Euler integration is implemented for the momentum equations, where its discretized form is as follows

$$\hat{\boldsymbol{u}} = \boldsymbol{u}^n + \Delta t(\nu\Delta\boldsymbol{u}^n - \nabla.(\boldsymbol{u}^n)^2). \tag{5.13}$$

Substituting the foregoing equation in Equation (5.12) results in

$$\boldsymbol{u}^{n+1} = \boldsymbol{u}^n + \Delta t(\nu\Delta\boldsymbol{u}^n - \nabla.(\boldsymbol{u}^n)^2 - \frac{1}{\rho}\nabla\mathcal{P}). \tag{5.14}$$

Using the physical pressure p instead of the pseudo-pressure $\mathcal{P}$ causes the integration has the accuracy of the forward Euler (McDonough, 2007). Regarding the numerical analysis performed by Kim and Moin (1985) one can find that

$$p = \mathcal{P} + \mathcal{O}(\Delta t/\text{Re}).$$

The difference in the accuracy of the pressure and pseudo-pressure functions is of the order $\mathcal{O}(\Delta t)$ and as a result this substitution generates a local error of the order $\mathcal{O}(\Delta t^2)$ resulting in a general error of the order $\mathcal{O}(\Delta t)$. It should also be noted that this is separate from the error caused by numerical integration. The decomposition of the equations in the fractional step method is the source of this error that cannot be easily resolved by increasing the accuracy of time integration methods. However, the projection method developed on this basis has first-order temporal accuracy. Finally, in the cases in which an accurate physical pressure field is needed, such as free-surface flows, it is necessary to solve the real pressure Poisson's equation. Although the $\mathcal{P} \sim p$ approximation has a good accuracy, in the case of free-surface flows and fluid–structure interaction, particularly where structures are sensitive to applied forces, it is necessary to solve the pressure Poisson's equation

$$\Delta p = -2(u_y v_x - u_x v_y), \tag{5.15}$$

with the following boundary conditions

$$
\begin{aligned}
p_x &= -(u_t + (u^2)_x + (uv)_y - \nu \Delta u) && \text{on vertical boundaries,} \\
p_y &= -(v_t + (uv)_x + (v^2)_y - \nu \Delta v) && \text{on horizontal boundaries.}
\end{aligned}
\tag{5.16}
$$

In cases where no physical pressure is required during the solution process, one can compute the pressure field at any time by solving Equation (5.15), since the exact value of the physical pressure is not necessary for the time evolution of the projection method (McDonough, 2007).

5.4 The Main Framework of the Projection Method

Since the development of the projection method, various types of it have been discussed; according to a paper by Pozrikidis (2003), all projection methods can be categorized into two main groups. The first group is devoted to methods based on pressure correction. In this group, the pressure term of momentum equations is approximated by extrapolation of the present pressure field. So the velocity field in the next time step is computed by integrating the momentum equations; then in the projection step, approximated pressure changes in the equations is corrected. In the second group, called the pure projection method, the pressure term is first removed from the momentum equations. Then by integrating the equations, the velocity field is obtained for the next time step. Since this velocity field is not solenoidal, corrections are imposed by continuity condition in the projection step. In this book we use the pure projection method.

5.4.1 Implementation of the Projection Method

The main steps for the numerical simulation with the projection method in a sample study are illustrated by the following procedure (McDonough, 2007)

1. Solving the Burgers' equations derived from the momentum equations using the following steps,
   (a) time integration using the fourth-order Runge–Kutta method;
   (b) spatial discretization of the terms on the right hand side of the equations;
   (c) spatial discretization of the terms on the left hand side of the equations;
   (d) applying the boundary conditions;
2. filtering the computed auxiliary velocity field ($\hat{\mathbf{u}}$) obtained from solving the Burgers' equations, if necessary;
3. implementation of the projection process in order to compute the physical velocity field for the next time step, that is $\mathbf{u}^{n+1}$, during the following steps,
   (a) taking the divergence of the auxiliary velocity field $\hat{\mathbf{u}}$,
   (b) solving the pseudo-pressure Poisson's equation,
   (c) using the pseudo-pressure field P to compute the velocity field for the next time step, that is $\mathbf{u}^{n+1}$,
4. solving the physical pressure Poisson's equation;
5. preparing the results for the post-processing.

These steps are illustrated in Figure 5.1. The main topic of this book is to apply the Krylov subspace solver to solve the system of equations obtained by discretizing the pseudo-pressure Poisson's equation, which is distinguished in Figure 5.1. In fact, the most time-consuming and costly steps of the projection method is to solve the pseudo-pressure Poisson's equation. By applying the Krylov subspace solvers, the speed-up factor is expected to increase dramatically.

5.4.2 Discretization of the Governing Equations

In order to numerically simulate the unsteady incompressible fluid flows, one can begin with momentum equation, Equation (5.6). Here, for the sake of simplicity of the equations, the terms representing the acceleration due to body forces are not taken into account. First by removing the pressure gradient term, the momentum equation turns into viscous Burgers' equation for auxiliary velocity field, that is

$$\hat{\mathbf{u}}_t = -\hat{\mathbf{u}}.\nabla\hat{\mathbf{u}} + \nu\nabla^2\hat{\mathbf{u}}. \tag{5.17}$$

To explain the discretization process of Equation (5.17), it can be represented in component form as

$$\begin{aligned}
\hat{u}_t &= -\hat{u}\hat{u}_x - \hat{v}\hat{u}_y + \nu(\hat{u}_{xx} + \hat{u}_{yy}), \\
\hat{v}_t &= -\hat{u}\hat{v}_x - \hat{v}\hat{v}_y + \nu(\hat{v}_{xx} + \hat{v}_{yy}).
\end{aligned} \tag{5.18}$$

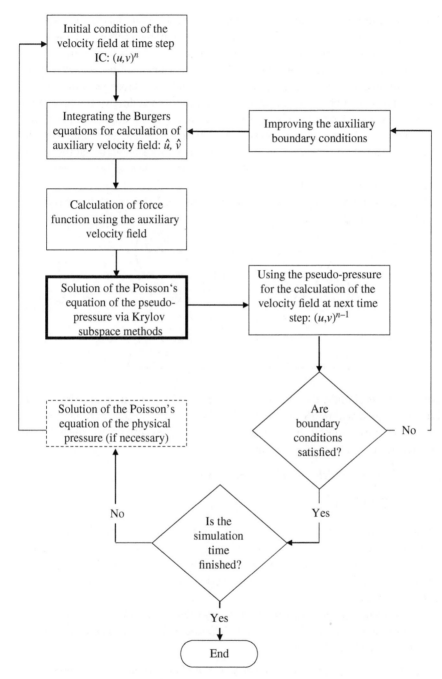

Figure 5.1 Chorin's projection method procedure.

Boundary and initial conditions are needed to solve Equations (5.18). Since the auxiliary velocity field does not have an initial value in the solution and due to the slight difference between the physical and the auxiliary velocity field, especially for the zones far from boundaries, the physical velocity field can be used as a guess for the initial conditions of the auxiliary velocity field. The error generated by taking this guess will be corrected during the solution procedure of the projection method. By considering the physical velocity boundary conditions and applying the appropriate boundary conditions for the auxiliary velocity field, initial conditions of the auxiliary velocity field will be corrected. More details about the procedure of applying the different boundary conditions for the auxiliary velocity field will be given later in this chapter.

The right hand side terms of Equations (5.18) can be discretized using the central finite difference approximation as

$$
\begin{aligned}
\frac{\partial \hat{u}^n}{\partial t} &= -\hat{u}^n_{i,j}\left(\frac{\hat{u}^n_{i+1,j} - \hat{u}^n_{i-1,j}}{2\Delta x}\right) - \hat{v}^n_{i,j}\left(\frac{\hat{u}^n_{i,j+1} - \hat{u}^n_{i,j-1}}{2\Delta y}\right) + \\
&\quad v\left(\frac{\hat{u}^n_{i+1,j} - 2\hat{u}^n_{i,j} + \hat{u}^n_{i-1,j}}{(\Delta x)^2} + \frac{\hat{u}^n_{i,j+1} - 2\hat{u}^n_{i,j} + \hat{u}^n_{i,j-1}}{(\Delta y)^2}\right) \\
\frac{\partial \hat{v}^n}{\partial t} &= -\hat{u}^n_{i,j}\left(\frac{\hat{v}^n_{i+1,j} - \hat{v}^n_{i-1,j}}{2\Delta x}\right) - \hat{v}^n_{i,j}\left(\frac{\hat{v}^n_{i,j+1} - \hat{v}^n_{i,j-1}}{2\Delta y}\right) + \\
&\quad v\left(\frac{\hat{v}^n_{i+1,j} - 2\hat{v}^n_{i,j} + \hat{v}^n_{i-1,j}}{(\Delta x)^2} + \frac{\hat{v}^n_{i,j+1} - 2\hat{v}^n_{i,j} + \hat{v}^n_{i,j-1}}{(\Delta y)^2}\right).
\end{aligned}
\tag{5.19}
$$

Henceforth, to simplify the equations, the right hand side terms of Equations (5.19) are represented with RHS_x and RHS_y. For the time integration of the Equations (5.19), the fourth-order Runge–Kutta method is used. By discretizing the left hand side terms of Equations (5.19) we have

$$
\begin{aligned}
\frac{\hat{u}^n - u^n}{\Delta t} &= RHS_x \quad \to \quad \hat{u}^n = \Delta t RHS_x + u^n \\
\frac{\hat{v}^n - v^n}{\Delta t} &= RHS_y \quad \to \quad \hat{v}^n = \Delta t RHS_y + v^n.
\end{aligned}
\tag{5.20}
$$

To simplify the explanation of the solution procedure, first the forward Euler method is considered as the time integration method. After the complete description of the velocity derivation procedure, implementation of a more sophisticated and accurate integration method, i.e. fourth-order Runge–Kutta, will be presented.

According to Equations (5.20), the auxiliary velocity field $(\hat{u}, \hat{v})$ is obtained in an integration step. Then, to derive a solenoidal velocity field, three steps of projection method should be implemented

1. taking the divergence of the auxiliary velocity field to construct the force function of the pseudo-pressure Poisson's equation,

2. solving the elliptic equation to compute the pseudo-pressure field $\mathcal{P}$,
3. computing the physical velocity field for the next time step, i.e. $\boldsymbol{u}^{n+1}$.

According to the mathematical theory mentioned for the projection method, after decomposing the momentum equations and computation of the auxiliary velocity field, by ignoring the pressure gradient term, necessary considerations for applying the pressure gradient term effect on the physical velocity field should be taken into account.

The right hand side of the Poisson's Equation (5.11) can be discretized as

$$\rho \frac{\nabla \cdot \hat{\boldsymbol{u}}}{\Delta t} \simeq \frac{\rho}{\Delta t} \left(\frac{\hat{u}_{i+1,j}^n - \hat{u}_{i-1,j}^n}{2\Delta x} + \frac{\hat{v}_{i,j+1}^n - \hat{v}_{i,j-1}^n}{2\Delta y} \right). \tag{5.21}$$

Now, the Krylov subspace solvers, which have been discussed in detail in previous chapters, are used to solve this Poisson's equation, i.e. Equation (5.11). Given the ellipticity of the pseudo-pressure Poisson's equation, the boundary conditions must be clear on all boundaries. It should be noted that the Neumann boundary condition should always be applied on solid boundaries; moreover, in the cases where physical pressure values are known on inlet and outlet boundaries, the Dirichlet boundary condition could be established by assuming that the physical and pseudo pressures are approximately equal.

Poisson's equation of the pseudo-pressure and its boundary condition are defined as

$$\nabla^2 \mathcal{P} = \alpha \quad \text{on } \Omega, \tag{5.22}$$

$$\frac{\partial \mathcal{P}}{\partial \boldsymbol{n}} = 0 \quad \text{on } \partial \Omega, \tag{5.23}$$

where α is a force function computed from Equation (5.21) and $\boldsymbol{n}$ is a unit vector normal to the boundary $\partial \Omega$. If the pure Neumann boundary condition is applied on all boundaries, discretizing and constructing the system of equations results in a singular problem. Hence a particular approach should be adopted to solve it. By considering a two-dimensional domain and using a second-order central finite difference approximation, Equation (5.22) can be written as follows

$$\frac{\partial^2 \mathcal{P}}{\partial x^2} + \frac{\partial^2 \mathcal{P}}{\partial y^2} = \alpha \rightarrow$$

$$\frac{\mathcal{P}_{i+1,j} - 2\mathcal{P}_{i,j} + \mathcal{P}_{i-1,j}}{(\Delta x)^2} + \frac{\mathcal{P}_{i,j+1} - 2\mathcal{P}_{i,j} + \mathcal{P}_{i,j-1}}{(\Delta y)^2} = \alpha_{i,j}. \tag{5.24}$$

The above equation can be simplified by taking a regular Cartesian grid $\Delta x = \Delta y = h$ as follows

$$\mathcal{P}_{i,j} = \frac{1}{4}[\mathcal{P}_{i-1,j} + \mathcal{P}_{i+1,j} + \mathcal{P}_{i,j-1} + \mathcal{P}_{i,j+1} - h^2 \alpha_{i,j}]. \tag{5.25}$$

Using Equation (5.25), the new value of the pseudo-pressure function can be computed on the points within the domain Ω. The resulting system of equations are then solved by Krylov subspace methods and pseudo-pressure function is obtained over the whole domain. Before starting the iteration operation, given the pure Neumann boundary condition at all boundaries and the singularity of the sparse matrix of the equations system, it is necessary to use an anchor point to obtain a unique solution. It means to fix a certain point of the computational domain which ensures the convergence of the solution in the Krylov subspace methods; see (Van der Vorst, 2003) for further details. By computing the pseudo-pressure function over Ω, it is possible to derive $\boldsymbol{u}^{n+1}$ with the following equations

$$
\begin{aligned}
u^{n+1} &= \hat{u} - \frac{\Delta t}{\rho} \frac{\partial P}{\partial x} \\
v^{n+1} &= \hat{v} - \frac{\Delta t}{\rho} \frac{\partial P}{\partial y}.
\end{aligned}
\tag{5.26}
$$

After computing the velocity field at the next time step and before advancing in time, it is necessary to correct the initial guess used for the auxiliary velocity field. Due to the specificity of the velocity field boundary conditions, the auxiliary velocity values on the boundaries can be modified using the following equations

$$
\begin{aligned}
\hat{u}_{BC} &= u_{BC} + \frac{\Delta t}{\rho} \frac{\partial P}{\partial x} \\
\hat{v}_{BC} &= v_{BC} + \frac{\Delta t}{\rho} \frac{\partial P}{\partial y}.
\end{aligned}
\tag{5.27}
$$

To assess the accuracy of the auxiliary velocity boundary conditions, the accuracy of the physical velocity boundary conditions can be examined. Thus, as long as the physical velocity accuracy criteria is not met, the auxiliary velocity field should be modified. After satisfying the boundary conditions of the physical velocity field and modifying the auxiliary velocity field, the solution can be advanced one time step further.

As mentioned before, a simple forward Euler integration method is applied to complete the description of the projection method procedure. Now, we introduce the application of the fourth-order Runge–Kutta integration method. In this method, the computation time accuracy will be improved by adding some auxiliary intermediate time points within the main time step. The fourth-order Runge–Kutta integration method has four computational phases in each time step, where all described procedures of the velocity computation should be followed in each phase. It should also be noted that there is no need to correct the auxiliary velocity boundary condition in all phases except the last one. The four phases of the fourth-order Runge–Kutta integration method for solving

Equation (5.18) are

$$\text{Phase1} \begin{cases} k_{1x} = \Delta t.RHS_x(u^n, v^n) \\ k_{1y} = \Delta t.RHS_y(u^n, v^n) \end{cases}$$

$$\text{Phase2} \begin{cases} k_{2x} = \Delta t.RHS_x(u^n + 0.5k_{1x}, v^n + 0.5k_{1y}) \\ k_{2y} = \Delta t.RHS_y(u^n + 0.5k_{1x}, v^n + 0.5k_{1y}) \end{cases}$$

$$\text{Phase3} \begin{cases} k_{3x} = \Delta t.RHS_x(u^n + 0.5k_{2x}, v^n + 0.5k_{2y}) \\ k_{3y} = \Delta t.RHS_y(u^n + 0.5k_{2x}, v^n + 0.5k_{2y}) \end{cases} \quad (5.28)$$

$$\text{Phase4} \begin{cases} k_{4x} = \Delta t.RHS_x(u^n + k_{3x}, v^n + k_{3y}) \\ k_{4y} = \Delta t.RHS_y(u^n + k_{3x}, v^n + k_{3y}) \end{cases}$$

$$u^{n+1} \simeq u^n + \frac{1}{6}(k_1 + 2k_2 + 2k_3 + k_4).$$

5.5 Numerical Case Study

5.5.1 Vortex Shedding from Circular Cylinder

The availability of the numerical solution results and empirical experiments on the flow through the cylinder has made this a benchmark for the validation of the numerical methods. For Reynolds numbers less than 47, the flow structures maintain their symmetry behind the cylinder. But as the Reynolds number increases, the symmetry of the vortices formed behind the cylinder is broken and the vortices begin to oscillate. The oscillation frequency of these vortices increases with increasing the Reynolds number and the flow structures form the Kármán vortex street phenomenon.

In the present study, a two-dimensional flow is simulated over a circular cylinder with diameter $d = 0.05m$ in a $1.5m \times 0.75m$ channel. The inlet uniform velocity U_∞ is $1m/s$ and Reynolds number is defined as $U_\infty d/v$; see Figure 5.2. The equations are solved on a regular Cartesian grid 512×256 and circular cylinder affects the flow field as an immersed boundary with no-slip boundary conditions. To validate the results, the dimensionless Strouhal number

$$St = f\frac{d}{U_\infty} \quad (5.29)$$

is used, where f is the oscillation frequency of the vortices formed behind the cylinder. Figure 5.3 shows the snapshots of the vorticity field around the circular cylinder at different time points. Table 5.1 presents the Strouhal number resulting from the present study and the other references. Due to the considerable distance of the

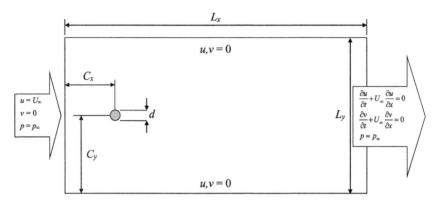

Figure 5.2 Configuration and boundary conditions in the problem of the vortex shedding from the circular cylinder. $C_x = 0.246m$, $C_y = 0.375m$, $L_x = 1.5m$, $L_y = 0.75m$, $d = 0.05m$ and $U_\infty = 1m/s$.

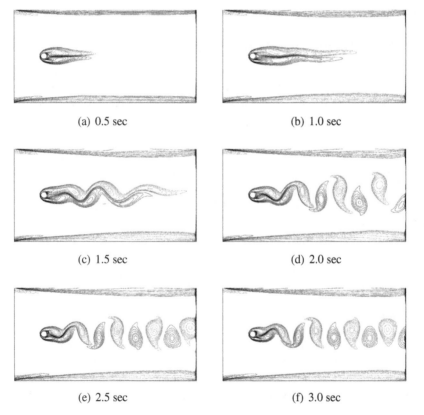

(a) 0.5 sec

(b) 1.0 sec

(c) 1.5 sec

(d) 2.0 sec

(e) 2.5 sec

(f) 3.0 sec

Figure 5.3 Six snapshots of the vorticity field around a fixed circular cylinder inside a channel. Negative contour levels are shown with dashed lines.

Table 5.1 Strouhal number resulting from the present study and the other references.

References	St
Present Study	0.167
Silva et al. (2003)	0.16
Williamson (1996)	0.166
Roshko (1955)	0.16
Lai and Peskin (2000)	0.165

outlet boundary from the cylinder location, the results show that accurately satisfying the outlet boundary conditions is unimportant. As mentioned in Chapter 1, the main purpose of this book is to introduce a package of Krylov subspace solvers that can be used in a CFD code. Here, the BiCGSTAB solver was used to solve the pseudo-pressure Poisson's equation. The major part of the computational cost in the projection method imposed by the pseudo-pressure Poisson's equation. Since, the time consuming solution of the Poisson's equation is repeated in each time step, choosing an efficient solver can play a significant role in dropping down the total CPU-time. The results show that using a stationary solver like SOR increases the total time of the numerical simulation up to three times that of the BiCGSTAB solver. The speed-up factors of the Krylov subspace methods can be investigated in an independent study.

5.5.2 Vortex Shedding from a Four-Leaf Cylinder

In this numerical case, flow over a four-leaf cylinder is simulated using the projection method. The geometrical arrangement of the problem is shown in Figure 5.4. The cross section geometry of the four-leaf cylinder is introduced using the polar equations as follows

$$r = A + B\cos(4\theta)$$
$$x = r\cos(\theta) \tag{5.30}$$
$$y = r\sin(\theta)$$

where $A = 0.04m$ and $B = 0.0125m$. Figure 5.5 shows the snapshots of the vorticity field around the four-leaf cylinder at different time points. It should be noted that

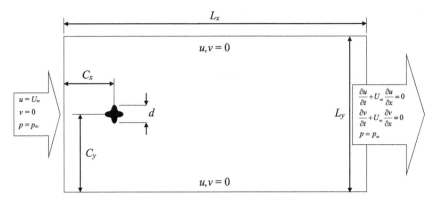

Figure 5.4 Configuration and boundary conditions in the problem of the vortex shedding from the four-leaf cylinder. $C_x = 0.246m$, $C_y = 0.375m$, $L_x = 1.5m$, $L_y = 0.75m$, $d = 0.105m$, $U_\infty = 1m/s$ and Re = 210.

the four-leaf cylinder affects the flow field as an immersed boundary with no-slip boundary conditions.

5.5.3 Oscillating Cylinder in Quiescent Fluid

In this subsection, the vortex production phenomenon due to the oscillatory motion of a cylinder in quiescent fluid is studied. Two main parameters which specify the flow characteristics are Reynolds number, $Re = U_{max}d/v$, and Keulegan–Carpenter number, $KC = U_{max}/(fd)$, in which U_{max} is the maximum velocity of the cylinder, d is the cylinder diameter, v is the kinematic viscosity and f is the linear oscillation frequency of the cylinder. The oscillation of the cylinder is defined with a simple periodic relation, i.e. $x_c(t) = -A \sin(2\pi ft)$, where $x_c(t)$ is the position of the cylinder center and A is the amplitude of the oscillation. This case is simulated in a $20d \times 20d$ rectangular domain with 400×400 computational grid. The Reynolds and Keulegan–Carpenter numbers are set to 100 and 5, respectively. The vorticity field in three different oscillation phases are presented in Figure 5.6. Figure 5.7 compares the present results of the horizontal and the vertical velocity in four different sections for the oscillation phases of 180°, 210° and 330°, with reference Liao et al. (2010). In this simulation, the BiCGSTAB solver is also used to solve the pseudo-pressure Poisson's equation.

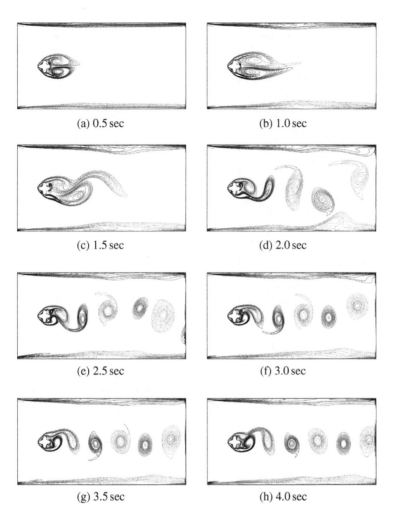

Figure 5.5 Eight snapshots of the vorticity field around a fixed four-leaf cylinder inside a channel. Negative contour levels are shown with dashed lines.

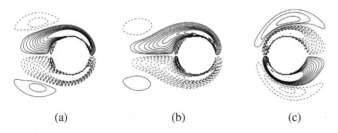

Figure 5.6 Vorticity field around an oscillating cylinder in three oscillation phases (a) 180°, (b) 210° and (c) 330°. Negative contour levels are shown with dashed lines.

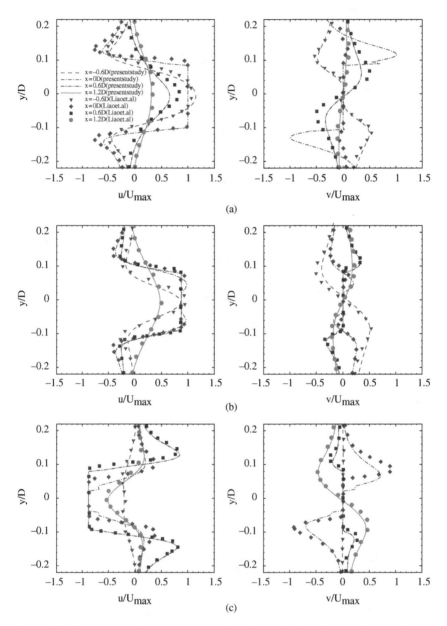

Figure 5.7 Comparison of the horizontal and vertical velocity in four sections for three oscillation phases (a) 180°, (b) 210°, (c) 330°, between the present study and reference Liao et al. (2010).

Exercises

5.1 Consider the vorticity-stream function formulation of the Navier–Stokes equations. In this secondary variables formulation we encounter stream function and pressure Poisson's equations as follows

$$\Delta \psi = -\omega,$$

$$\Delta p = 2\rho \left(\frac{\partial^2 \psi}{\partial x^2} \frac{\partial^2 \psi}{\partial y^2} - \left(\frac{\partial^2 \psi}{\partial x \partial y} \right)^2 \right).$$

In the structure of the vorticity-stream function formulation, solving the pressure Poisson's equation is not necessary because the formulation is essentially derived by eliminating the pressure gradient terms. So this equation is solved when we want to compute the pressure field at a particular time step. Applying the efficient elliptic solvers has a remarkable role in reducing the computational time and error. The incompressible flow in a cavity is a benchmark problem that has been addressed in most CFD textbooks. Consider a square $[0, 1] \times [0, 1]$ lid-driven cavity with a top velocity of $1m/s$ while the rest of the sides have a no-slip boundary condition.

(a) Write the algorithm of the vorticity-stream function formulation for the solution of the incompressible Navier–Stokes equations and show where the Krylov subspace methods can be applied as the solvers of the elliptic part of the equations.

(b) Run the developed program of Appendix G and evaluate the computational time of the solution precess when the Krylov subspace methods of Appendix B and Appendix C are applied as the elliptic part solver of the formulation. For all methods, compare the computational time variations when we need 10, 20, and 30 seconds of the solution, respectively.

5.2 Run the program of Appendix G and evaluate the effect of matrix–vector multiplication algorithm on computational time of the solution when we need 10, 20 and 30 seconds of the solution, respectively.

5.3 Repeat the Exercise 2 when the grid size is halved or doubled.

5.4 Run the program of Appendix G and evaluate the effect of Reynolds number on the computational time of the solution, when the CG method with various formats of matrix–vector multiplication is applied as the solver of the stream function Poisson's equation.

5.5 Run the program of Appendix G when the GMRES∗ is applied as the solver of the stream function Poisson's equation and evaluate the computational time of the solution when the number of inner iterations of the preconditioners changes.

5.6 Run the program of Appendix G when the CG and ILU(0)-CG methods are applied as the solver of the stream function Poisson's equation. Compare the computational time of the solution from low to high grid resolutions when the CSR format is used for the matrix–vector multiplication.

5.7 Develop a subroutine with FORTRAN or C language to solve the pressure Poisson's equation introduced in Exercise 1. Apply the CG, BiCG, CGS and BiCGSTAB methods with their preconditioned versions as the Poisson's equation solver and compare their convergence trend and CPU-time.

5.8 Develop a FORTRAN or C program based on the projection algorithm discussed in Chapter 5 and solve the benchmark problem of the Exercise 1. Use the developed subroutines of Appendix B and Appendix C as the solvers of the pseudo-pressure Poisson's equation.

(a) Apply the BiCG, CGS, BiCGSTAB methods and their preconditioned versions and report the computational time of the solution when we need 10, 20 and 30 seconds of the solution, respectively.

(b) Evaluate the effect of matrix–vector multiplication algorithm on the computational time of the projection method.

5.9 Consider the program of Appendix G and try to extract the algorithm of the subroutines UNSTRUCT_GRID, NEIGHBORS, RHS_CONSTITUTION and MATRIX_CONSTITUTION.

Appendix A

Sparse Matrices

A.1 Storing the Sparse Matrices

A.1.1 Coordinate to CSR Format Conversion

This subroutine converts a stored matrix in coordinate format into CSR format. On entry, NROW denotes the dimension of the matrix and NNZ is the number of non-zero elements in matrix. The matrix in the coordinate format is defined using three vectors A, IR and JC. The real vector A stores the actual real value of the NNZ non-zero elements. IR and JC store the row and column number of the corresponding non-zero elements, respectively. It should be noted that the order of the elements is arbitrary.

On return, the vector IR is destroyed. AO and JAO contain the non-zero real values and their column indices, respectively. IAO is the pointer to the beginning of the row, in arrays AO and JAO.

```
1  !****************************************************
2  !          Coordinate to CSR Conversion
3  !****************************************************
4  SUBROUTINE COOCSR (NROW,NNZ,A,IR,JC,AO,JAO,IAO)
5  !--------------------------------------------------
6  REAL(8),DIMENSION(NNZ)           :: A,AO
7  INTEGER,DIMENSION(NNZ)           :: IR,JC,JAO
8  INTEGER,DIMENSION(NROW+1)        :: IAO
9  REAL(8)                          :: X
10 !--------------------------------------------------
11 DO K=1,NROW+1
12    IAO(K)=0
13 END DO
14 !--------------------------------------------------
15 ! determine row-lengths.
16 !--------------------------------------------------
17 DO K=1,NNZ
18    IAO(IR(K))=IAO(IR(K))+1
19 END DO
```

Krylov Subspace Methods with Application in Incompressible Fluid Flow Solvers, First Edition.
Iman Farahbakhsh.
© 2020 John Wiley & Sons Ltd. Published 2020 by John Wiley & Sons Ltd.
Companion Website: www.wiley.com/go/Farahbakhs/KrylovSubspaceMethods

```
20  !-------------------------------------------------
21  ! starting position of each row.
22  !-------------------------------------------------
23  K=1
24  DO  J=1,NROW+1
25      K0=IAO(J)
26      IAO(J)=K
27      K=K+K0
28  END DO
29  !-------------------------------------------------
30  ! go through the structure  once more. Fill in output matrix.
31  !-------------------------------------------------
32  DO  K=1,NNZ
33      I=IR(K)
34      J=JC(K)
35      X=A(K)
36      IAD=IAO(I)
37      AO(IAD)=X
38      JAO(IAD)=J
39      IAO(I)=IAD+1
40  END DO
41  !-------------------------------------------------
42  ! shift back IAO
43  !-------------------------------------------------
44  DO  J=NROW,1,-1
45      IAO(J+1)=IAO(J)
46  END DO
47  IAO(1)=1
48  !-------------------------------------------------
49  END SUBROUTINE COOCSR
```

A.1.2   CSR to MSR Format Conversion

This subroutine converts a stored matrix in CSR format into MSR format. On entry, TL denotes the NNZ+1 and as mentioned in Appendix A.1.1, NROW and NNZ are the matrix dimension and the number of non-zero elements, respectively. In this subroutine, AA, JA and IA are the same as AO, JAO and IAO on return of the previous subroutine of Appendix A.1.1, respectively.

On return, two vectors AO and JAO define the sparse matrix in MSR format. AO(1:N) contains the diagonal of the matrix. AO(N+2:TL) contains the non-diagonal elements of the matrix, stored row-wise. JAO(1:N+1) contains the pointer array for the non-diagonal elements in AO(N+1:TL) and JAO(N+2:TL), i.e., for I less than N+1, JAO(I) points to beginning of row I in arrays AO, and JAO. JAO(N+2:TL) contains the column indices of non-diagonal elements.

In this subroutine two auxiliary vectors, i.e., a real work array WK(NROW) and an integer work array IWK(NROW+1) are used to complete the process of constructing the vectors AO(TL) and JAO(TL).

```
1  !*****************************************************
2  !             CSR to MSR Conversion
3  !*****************************************************
4  SUBROUTINE CSRMSR   (TL,NROW,NNZ,AA,JA,IA,AO,JAO)
5  !---------------------------------------------------
6  IMPLICIT NONE
7  INTEGER                           :: I,J,K,NNZ,II,TL,NROW
8  REAL(8),DIMENSION(NNZ)            :: AA
9  REAL(8),DIMENSION(TL)             :: AO
10 REAL(8),DIMENSION(NROW)           :: WK
11 INTEGER,DIMENSION(NROW+1)         :: IA,IWK
12 INTEGER,DIMENSION(NNZ)            :: JA
13 INTEGER,DIMENSION(TL)             :: JAO
14 !---------------------------------------------------
15 DO  I=1,NROW
16     WK(I) = 0.D0
17     IWK(I+1) = IA(I+1)-IA(I)
18     DO  K=IA(I),IA(I+1)-1
19         IF (JA(K).EQ.I) THEN
20             WK(I)= AA(K)
21             IWK(I+1)=IWK(I+1)-1
22         END IF
23     END DO
24 END DO
25 !---------------------------------------------------
26 ! Copy backwards (to avoid collisions)
27 !---------------------------------------------------
28 DO II=NROW,1,-1
29     DO K=IA(II+1)-1,IA(II),-1
30        J=JA(K)
31        IF (J.NE.II) THEN
32            AO(TL) = AA(K)
33            JAO(TL) = J
34            TL = TL-1
35        END IF
36     END DO
37 END DO
38 !---------------------------------------------------
39 ! Compute pointer values and copy WK(NROW)
40 !---------------------------------------------------
41 JAO(1)=NROW+2
42 DO  I=1,NROW
43     AO(I)=WK(I)
44     JAO(I+1)=JAO(I)+IWK(I+1)
45 END DO
46 !---------------------------------------------------
47 END SUBROUTINE CSRMSR
```

A.1.3   CSR to Ellpack-Itpack Format Conversion

This subroutine converts a stored matrix in CSR format into Ellpack-Itpack format. On entry, NROW is the matrix dimension and NNZ denotes the number of non-zero elements. Moreover, vectors A, IA and JA construct the sparse matrix in CSR format. NCOEF is the first dimension of arrays COEF, and JCOEF. MAXCOL is an integer equal to the number of columns available in COEF.

On return, `COEF` is a real array containing the values of the matrix A in Ellpack-Itpack format. This format is completed with an integer array containing the column indices of `COEF(i,j)` in A. The number of active diagonals which is found is stored in `NDIAG`. `IERR` is an error message which is 0 for correct return. If `IERR.NE.0` on return, this means that the number of diagonals, i.e., `NDIAG` exceeds `MAXCOL`.

```fortran
1  !****************************************************
2  !        CSR to Ellpack-Itpack Conversion
3  !****************************************************
4  SUBROUTINE CSRELL (NROW,NN,A,JA,IA,MAXCOL,COEF,JCOEF,NCOEF,NDIAG,IERR)
5  IMPLICIT NONE
6  INTEGER              :: NROW,NN,NCOEF,NDIAG,IERR,I,J,K,K1,K2,MAXCOL
7  INTEGER,DIMENSION(NROW+1)      :: IA
8  INTEGER,DIMENSION(NN)          :: JA
9  INTEGER,DIMENSION(NCOEF,NDIAG) :: JCOEF
10 REAL(8),DIMENSION(NN)          :: A
11 REAL(8),DIMENSION(NCOEF,NDIAG) :: COEF
12 !------------------------------------------------
13 ! first determine the length of each row of lower-part-of(A)
14 !------------------------------------------------
15 IERR=0
16 NDIAG=0
17 DO  I=1,NROW
18     K=IA(I+1)-IA(I)
19     NDIAG=MAX0(NDIAG,K)
20 END DO
21 !------------------------------------------------
22 ! check whether sufficient columns are available.
23 !------------------------------------------------
24 IF (NDIAG.GT.MAXCOL) THEN
25     IERR=1
26 ENDIF
27 !------------------------------------------------
28 ! fill COEF with zero elements and jcoef with row numbers.
29 !------------------------------------------------
30 DO  J=1,NDIAG
31     DO  I=1,NROW
32         COEF(I,J)=0.D0
33         JCOEF(I,J)=I
34     END DO
35 END DO
36 !------------------------------------------------
37 !        copy elements row by row.
38 !------------------------------------------------
39 DO  I=1,NROW
40     K1=IA(I)
41     K2=IA(I+1)-1
42     DO  K=K1,K2
43         COEF(I,K-K1+1)=A(K)
44         JCOEF(I,K-K1+1)=JA(K)
45     END DO
46 END DO
47 !------------------------------------------------
48 END SUBROUTINE CSRELL
```

A.1.4 CSR to Diagonal Format Conversion

This subroutine extracts IDIAG diagonals from the input matrix A, JA and IA, and puts the rest of the matrix in the output matrix AO, JAO and IAO. The diagonals to be extracted depend on the value of JOB. In the first case, the diagonals to be extracted are simply identified by their offsets provided in IOFF by the caller. In the second case, the code internally determines the IDIAG most significant diagonals, i.e., those diagonals of the matrix which have the largest number of the non-zero elements, and extracts them.

On entry, MD and NNZ denote the matrix dimension and the number of the non-zero elements of the matrix, respectively. IDIAG is an integer equal to the number of diagonals to be extracted. It should be noted that IDIAG may be modified on return. The real vector A and integer vectors JA and IA construct the matrix in CSR format. JOB serves as an indicator and is better thought of as a two-digit number JOB=xy. If the first digit (x) is one on entry then the diagonals to be extracted are internally determined. In this case subroutine CSRDIA extracts the IDIAG most important diagonals, i.e., those having the largest number on non-zero elements. If the first digit is zero then CSRDIA assumes that IOFF contains the offsets of the diagonals to be extracted. There is no verification that IOFF contains valid entries. The second digit (y) of JOB determines whether or not the remainder of the matrix is to be written on AO, JAO and IAO. If it is zero then AO, JAO and IAO is not filled, i.e., the diagonals are found and put in array DIAG and the rest is discarded. If it is one, AO, JAO and IAO contains matrix of the remaining elements. Thus JOB=0 means do not select diagonals internally (pick those defined by IOFF) and do not fill AO, JAO and IAO. JOB=1 means do not select diagonals internally and fill AO, JAO and IAO. JOB=10 means select diagonals internally and do not fill AO, JAO and IAO. JOB=11 means select diagonals internally and fill AO, JAO and IAO. NDIAG denotes an integer equal to the first dimension of array DIAG.

On return, IDIAG is the number of found diagonals. This may be smaller than its value on entry. DIAG is a real array of size (NDIAG, IDIAG) containing the diagonals of A on return. IOFF is an integer array of length IDIAG, containing the offsets of the diagonals to be extracted. The real vector AO and integer vectors JAO and IAO are remainders of the matrix in CSR format. IND is an integer array of length 2*MD-1 and is used as an integer work space. It is only needed when JOB.GE.10, i.e., in the case that the diagonals are to be selected internally.

It should be noted that, the algorithm is efficient if AO, JAO and IAO can be overwritten on A, JA and IA. When the code is required to select the diagonals (JOB.GE.10), the selection of the diagonals is done from left to right. If several diagonals have the same weight (number of non-zero elements) the leftmost one is selected first.

```
1    !*****************************************************
2    !          CSR to Diagonal Conversion
3    !*****************************************************
4    SUBROUTINE CSRDIA (MD,NNZ,IDIAG,JOB,A,JA,IA,NDIAG,DIAG,IOFF,AO,JAO,IAO,
         IND)
5    !----------------------------------------------------
6    IMPLICIT NONE
7    INTEGER                         :: MD,NNZ,IDIAG,NDIAG,JOB1,JOB2,JOB,N2,
         IDUM,II,JMAX,K,I,J,L,KO
8    REAL(8),DIMENSION(NDIAG,IDIAG)  :: DIAG
9    REAL(8),DIMENSION(NNZ)          :: A,AO
10   INTEGER,DIMENSION(NNZ)          :: JA,JAO
11   INTEGER,DIMENSION(MD+1)         :: IA,IAO
12   INTEGER,DIMENSION(2*MD-1)       :: IND
13   INTEGER,DIMENSION(IDIAG)        :: IOFF
14   !----------------------------------------------------
15   JOB1 = JOB/10
16   JOB2 = JOB-JOB1*10
17   IF (JOB1 .NE. 0) THEN
18       N2 = MD+MD-1
19       CALL INFDIA(MD,NNZ,JA,IA,IND,IDUM)
20   !----------------------------------------------------
21   ! determine diagonals to  accept.
22   !----------------------------------------------------
23       II=0
24       DO
25           II=II+1
26           JMAX=0
27           DO   K=1,N2
28               J=IND(K)
29               IF (JMAX.LT.J) THEN
30                   I=K
31                   JMAX=J
32               END IF
33           END DO
34           IF (JMAX.LE.0) THEN
35               II=II-1
36               EXIT
37           ENDIF
38           IOFF(II)=I-MD
39           IND(I)=-JMAX
40           IF (IDIAG.LE.II) THEN
41               EXIT
42           END IF
43       END DO
44       IDIAG=II
45   END IF
46   !----------------------------------------------------
47   ! initialize diago to zero
48   !----------------------------------------------------
49   DO  J=1,IDIAG
50       DO  I=1,MD
51           DIAG(I,J)=0.D0
52       END DO
53   END DO
54   KO=1
55   !----------------------------------------------------
56   ! extract diagonals and accumulate remaining matrix.
57   !----------------------------------------------------
58   DO   I=1,MD
```

```
59      DO  K=IA(I),IA(I+1)-1
60          J=JA(K)
61          DO L=1,IDIAG
62              IF (J-I.EQ.IOFF(L)) THEN
63                  DIAG(I,L)=A(K)
64                  GOTO 51
65              END IF
66          END DO
67  !-------------------------------------------------------
68  ! append element not in any diagonal to AO,JAO,IAO
69  !-------------------------------------------------------
70          IF (JOB2.NE.0) THEN
71              AO(KO)=A(K)
72              JAO(KO)=J
73              KO=KO+1
74          END IF
75  51      CONTINUE
76      END DO
77          IF (JOB2.NE.0) THEN
78              IND(I+1)=KO
79          END IF
80  END DO
81  !-------------------------------------------------------
82  !       finish with IAO
83  !-------------------------------------------------------
84  IF (JOB2.NE.0) THEN
85      IAO(1)=1
86      DO I=2,MD+1
87          IAO(I)=IND(I)
88      END DO
89  END IF
90  !-------------------------------------------------------
91  END SUBROUTINE CSRDIA
```

In the subroutine CSRDIA, the subroutine INFDIA is called which can obtain the information on the diagonals of the matrix A. This subroutine finds the lengths of each of the $2*N-1$ diagonals of A and it also returns the number of non-zero diagonals.

On entry, MD is the dimension of the matrix and NNZ denotes the number of the non-zero elements. In this subroutine, the first real vector of CSR format is no longer needed and the integer vectors JA and IA are used.

On return, IDIAG denotes the number of non-zero diagonals and IND is an integer array of length at least $2*MD-1$. The Kth entry in IND contains the number of non-zero elements in the diagonal number K. The numbering starts from the lowermost diagonal, i.e, bottom-left diagonal. In other words IND(K) is the length of a diagonal whose offset with respect to the main diagonal is K-MD.

```
1   !***************************************************
2   !                    INFDIA
3   !***************************************************
4   SUBROUTINE INFDIA (MD,NNZ,JA,IA,IND,IDIAG)
5   !---------------------------------------------------
6   IMPLICIT NONE
7   INTEGER                        :: MD,NNZ,N2,I,K,J,IDIAG
8   INTEGER,DIMENSION(MD+1)        :: IA
9   INTEGER,DIMENSION(2*MD-1)      :: IND
10  INTEGER,DIMENSION(NNZ)         :: JA
11  !---------------------------------------------------
12  N2=MD+MD-1
13  DO I=1,N2
```

```
14      IND(I)=0
15 END DO
16 DO I=1,MD
17    DO K=IA(I),IA(I+1)-1
18       J=JA(K)
19          IND(MD+J-I)=IND(MD+J-I)+1
20    END DO
21 END DO
22 !-------------------------------------------------
23 !     count the nonzero ones.
24 !-------------------------------------------------
25 IDIAG=0
26 DO K=1,N2
27    IF (IND(K).NE.0) IDIAG=IDIAG+1
28 END DO
29 !-------------------------------------------------
30 END SUBROUTINE INFDIA
```

A.2 Matrix-Vector Multiplication

A.2.1 CSR Format Matrix-Vector Multiplication

This subroutine multiplies a matrix by a vector using the dot product form, where the matrix is stored in the compressed sparse row format using three vectors AA_AUX, JA_AUX and IA_AUX where its details are previously presented in Section 2.4.2.2. On entry, MD and NNZERO denote the matrix dimension and number of non-zero elements. In addition to three mentioned vectors in the input list of subroutine, VE is considered as a real vector of length MD to be multiplied by the matrix in CSR format. On return, YY is a real vector of length MD, containing the product of the matrix in compressed form and vector VE.

```
1 !****************************************************
2 !    CSR Format Matrix-Vector Multiplication
3 !****************************************************
4 SUBROUTINE CSR_MAT_V_PRODUCT (MD,NNZERO,AA_AUX,JA_AUX,IA_AUX,VE,YY)
5 !-------------------------------------------------
6 IMPLICIT NONE
7 INTEGER                        :: I,M1,M2,NNZERO,MD
8 REAL(8),DIMENSION(MD)          :: VE,YY
9 REAL(8),DIMENSION(NNZERO)      :: AA_AUX
10 INTEGER,DIMENSION(MD+1)        :: IA_AUX
11 INTEGER,DIMENSION(NNZERO)      :: JA_AUX
12 !-------------------------------------------------
13 YY=0.D0
14 DO I=1,MD
15    M1=IA_AUX(I)
16    M2=IA_AUX(I+1)-1
17    YY(I)=DOT_PRODUCT(AA_AUX(M1:M2),VE(JA_AUX(M1:M2)))
18 END DO
19 !-------------------------------------------------
20 END SUBROUTINE CSR_MAT_V_PRODUCT
```

A.2.2   MSR Format Matrix-Vector Multiplication

This subroutine multiplies a matrix by a vector using the dot product form, where the matrix is stored in the MSR format using two vectors SA and IJA, where its details are previously presented in Section 2.4.2.5. On entry, TL and MD denote the NNZ+1 and matrix dimension, respectively; see Appendix A.1.2. As mentioned previously, the sparse matrix in MSR format is defined using two vectors of length TL. In the input list of the following subroutine the real vector X of length MD is multiplied with the matrix in the MSR format. On return, B is a real vector of length MD, containing the product of the matrix in compressed form and vector X.

```
 1 !*************************************************
 2 !    MSR Format Matrix-Vector Multiplication
 3 !*************************************************
 4 SUBROUTINE MSR_MAT_V_PRODUCT (TL,MD,SA,IJA,X,B)
 5 !-----------------------------------------------
 6 IMPLICIT NONE
 7 INTEGER                          :: I,K,TL,MD
 8 REAL(8),DIMENSION(MD)            :: B,X
 9 REAL(8),DIMENSION(TL)            :: SA
10 INTEGER,DIMENSION(TL)            :: IJA
11 !-----------------------------------------------
12 IF (IJA(1).NE.MD+2) PAUSE 'mismatched vector and matrix in sprsax'
13 DO I=1,MD
14    B(I)=SA(I)*X(I)
15    DO K=IJA(I),IJA(I+1)-1
16       B(I)=B(I)+SA(K)*X(IJA(K))
17    END DO
18 END DO
19 !-----------------------------------------------
20 END SUBROUTINE MSR_MAT_V_PRODUCT
```

A.2.3   Ellpack-Itpack Format Matrix-Vector Multiplication

This subroutine multiplies a matrix by a vector when the original matrix is stored in the Ellpack-Itpack format. On entry, N is the matrix dimension and X is a real array of the length N. The real and integer arrays of the Ellpack-Itpack format are A and JA. NA is the first dimension of the arrays A and JA as declared in the type declaration of the subroutine. The number of active columns in array A is determined by NCOL, i.e., the number of generalized diagonals in the original matrix. A(I,1:NCOL) contains the elements of row I in the original matrix and JA(I,1:NCOL) contains their column numbers.

```
 1 !*************************************************
 2 ! Ellpack-Itpack Format Matrix-Vector Multiplication
 3 !*************************************************
 4 SUBROUTINE ELLPACK_ITPACK_MAT_V_PRODUCT (N,X,Y,NA,NCOL,A,JA)
 5 !-----------------------------------------------
 6 IMPLICIT NONE
 7 INTEGER                          :: N,NA,NCOL,I,J
 8 REAL(8),DIMENSION(N)             :: X,Y
```

```
 9  REAL(8),DIMENSION(NA,NCOL)                    :: A
10  INTEGER,DIMENSION(NA,NCOL)                    :: JA
11  !---------------------------------------------------------
12  DO I=1,N
13      Y(I)=0.D0
14  END DO
15  DO J=1,NCOL
16      DO  I=1,N
17          Y(I)=Y(I)+A(I,J)*X(JA(I,J))
18      END DO
19  END DO
20  !---------------------------------------------------------
21  END SUBROUTINE ELLPACK_ITPACK_MAT_V_PRODUCT
```

A.2.4 Diagonal Format Matrix-Vector Multiplication

This subroutine multiplies a matrix by a vector when the original matrix is stored in the diagonal format. On entry, MD denotes the matrix dimension. X is a real array of length equal to the column dimension of the matrix. NDIAG is an integer number that shows the first dimension of array ADIAG as declared in the type declaration section of the subroutine. The number of diagonals in the matrix is determined using an integer scalar that is called IDIAG. DIAG is a real array of size (NDIAG, IDIAG) containing the diagonals of the matrix. DIAG(I, K) contains the element A(I, I+IOFF(K)) of the matrix, where, IOFF is an integer array of length IDIAG which shows the offsets of the matrix diagonals. On return, a real array of length MD, that is Y, contains the product A*X, where A is presented in diagonal format.

```
 1  !*****************************************************
 2  ! Diagonal Format Matrix-Vector Multiplication
 3  !*****************************************************
 4  SUBROUTINE DIAGONAL_MAT_V_PRODUCT (MD,X,Y,DIAG,NDIAG,IDIAG,IOFF)
 5  !---------------------------------------------------------
 6  IMPLICIT NONE
 7  INTEGER                          :: MD,J,K,IO,I1,I2,NDIAG,IDIAG
 8  INTEGER,DIMENSION(IDIAG)         :: IOFF
 9  REAL(8),DIMENSION(MD)            :: X,Y
10  REAL(8),DIMENSION(NDIAG,IDIAG)   :: DIAG
11  !---------------------------------------------------------
12  DO J=1,MD
13      Y(J)=0.D0
14  END DO
15  DO J=1,IDIAG
16      IO=IOFF(J)
17      I1=MAX0(1,1-IO)
18      I2=MIN0(MD,MD-IO)
19      DO K=I1,I2
20          Y(K)=Y(K)+DIAG(K,J)*X(K+IO)
21      END DO
22  END DO
23  !---------------------------------------------------------
24  END SUBROUTINE DIAGONAL_MAT_V_PRODUCT
```

A.3 Transpose Matrix-Vector Multiplication

A.3.1 CSR Format Transpose Matrix-Vector Multiplication

This subroutine multiplies the transpose of a matrix by a vector when the original matrix is stored in CSR format. This can also be viewed as the product of a matrix by a vector when the original matrix is stored in the compressed sparse column format. The argument list of the subroutine is the same as the subroutine CSR-MAT-V-PRODUCT; see Appendix A.2.1.

```
1  !*****************************************************
2  !CSR Format Transpose Matrix-Vector Multiplication
3  !*****************************************************
4  SUBROUTINE CSR_TMAT_V_PRODUCT (MD,NNZERO,AA_AUX,JA_AUX,IA_AUX,VE,YY)
5  !-------------------------------------------------
6  IMPLICIT NONE
7  INTEGER                        :: I,K,NNZERO,MD
8  REAL(8),DIMENSION(MD)          :: VE,YY
9  REAL(8),DIMENSION(NNZERO)      :: AA_AUX
10 INTEGER,DIMENSION(MD+1)        :: IA_AUX
11 INTEGER,DIMENSION(NNZERO)      :: JA_AUX
12 !-------------------------------------------------
13 DO I=1,MD
14    YY(I)=0.D0
15 END DO
16 !-------------------------------------------------
17 ! loop over the rows
18 !-------------------------------------------------
19 DO I=1,MD
20    DO K=IA_AUX(I),IA_AUX(I+1)-1
21       YY(JA_AUX(K))=YY(JA_AUX(K))+VE(I)*AA_AUX(K)
22    END DO
23 END DO
24 !-------------------------------------------------
25 END SUBROUTINE CSR_TMAT_V_PRODUCT
```

A.3.2 MSR Format Transpose Matrix-Vector Multiplication

This subroutine multiplies the transpose of a matrix by a vector when the original matrix is stored in MSR format. The argument list of the subroutine is the same as the subroutine MSR-MAT-V-PRODUCT; see Appendix A.2.2.

```
1  !*****************************************************
2  !MSR Format Transpose Matrix-Vector Multiplication
3  !*****************************************************
4  SUBROUTINE MSR_TMAT_V_PRODUCT (TL,MD,SA,IJA,X,B)
5  IMPLICIT NONE
6  INTEGER                        :: I,J,K,TL,MD
7  REAL(8),DIMENSION(MD)          :: B,X
8  INTEGER,DIMENSION(TL)          :: IJA
9  REAL(8),DIMENSION(TL)          :: SA
10 !-------------------------------------------------
11 IF (IJA(1).NE.MD+2) PAUSE 'mismatched vector and matrix in
       MSR_TMAT_V_PRODUCT'
```

```
12  DO I=1,MD
13      B(I)=SA(I)*X(I)
14  END DO
15  DO I=1,MD
16      DO K=IJA(I),IJA(I+1)-1
17          J=IJA(K)
18          B(J)=B(J)+SA(K)*X(I)
19      END DO
20  END DO
21  !-------------------------------------------------
22  END SUBROUTINE MSR_TMAT_V_PRODUCT
```

A.4   Matrix Pattern

To generate the matrix pattern and observe the distribution of the non-zero elements in sparse matrix, one can use the following subroutine. The subroutine
argument contains a list of integer scalar and vector inputs. JOB1 is an integer
flag the value of which points to a special matrix in Table 2.1. TNNZ and N denote
the number of non-zero elements and matrix dimension, respectively, and belong
to a special matrix which is pointed out by JOB1. IR and JC are integer vectors
that store the row and column indices of the pointed matrix with integer flag JOB1.
The output of the subroutine will be saved with . PLT extension.

```
1   !*****************************************************
2   !                   Matrix Pattern
3   !*****************************************************
4   SUBROUTINE MATRIX_PATTERN (JOB1,TNNZ,N,IR,JC)
5   !-------------------------------------------------
6   IMPLICIT NONE
7   INTEGER                     :: I,JOB1,N,TNNZ
8   INTEGER,DIMENSION(TNNZ)     :: JC,IR
9   !-------------------------------------------------
10  IF (JOB1.EQ.1) THEN
11          OPEN(UNIT=JOB1+100,FILE='plat1919.PLT')
12  ELSE IF (JOB1.EQ.2) THEN
13          OPEN(UNIT=JOB1+100,FILE='plat362.PLT')
14  ELSE IF (JOB1.EQ.3) THEN
15          OPEN(UNIT=JOB1+100,FILE='gr_30_30.PLT')
16  ELSE IF (JOB1.EQ.4) THEN
17          OPEN(UNIT=JOB1+100,FILE='bcsstk22.PLT')
18  ELSE IF (JOB1.EQ.5) THEN
19          OPEN(UNIT=JOB1+100,FILE='nos4.PLT')
20  ELSE IF (JOB1.EQ.6) THEN
21          OPEN(UNIT=JOB1+100,FILE='hor__131.PLT')
22  ELSE IF (JOB1.EQ.7) THEN
23          OPEN(UNIT=JOB1+100,FILE='nos6.PLT')
24  ELSE IF (JOB1.EQ.8) THEN
25          OPEN(UNIT=JOB1+100,FILE='nos7.PLT')
26  ELSE IF (JOB1.EQ.9) THEN
27          OPEN(UNIT=JOB1+100,FILE='orsirr_1.PLT')
28  ELSE IF (JOB1.EQ.10) THEN
29          OPEN(UNIT=JOB1+100,FILE='sherman4.PLT')
30  ELSE IF (JOB1.EQ.11) THEN
31          OPEN(UNIT=JOB1+100,FILE='orsirr_2.PLT')
32  ELSE IF (JOB1.EQ.12) THEN
33          OPEN(UNIT=JOB1+100,FILE='orsreg_1.PLT')
```

```
34 END IF
35 !------------------------------------------------
36 WRITE (JOB1+100,*) 'VARIABLES=I,J'
37 WRITE (JOB1+100,*) 'ZONE'
38 WRITE (JOB1+100,*) 'F=POINT'
39 DO I=1,TNNZ
40 WRITE (JOB1+100,*) JC(I),N-IR(I)+1
41 END DO
42 !------------------------------------------------
43 END SUBROUTINE MATRIX_PATTERN
```

Appendix B

Krylov Subspace Methods

B.1 Conjugate Gradient Method

This subroutine is written based on the Algorithm 3.3 and can solve a linear system and returns the solution vector. In the subroutine argument we have known parameters MD and NNZERO in the input list. The sparse matrix is stored in CSR format and presented via three vector AA_AUX, JA_AUX and IA_AUX in the input list of the subroutine argument. One can check the type and dimension of arrays in the type declaration section of the code. Two last real vectors X_OLD and R_OLD in the input list denote the initial guess and initial residual vector. The last variable of the argument list that is considered as the only output is the solution vector X. This input and output list is similar for all Krylov subspace solvers subroutines of Appendices B and C.

As mentioned in Section 2.4.3, the matrix–vector multiplication is the kernel of each Krylov subspace method which must be redefined in compressed format. In this subroutine four compressed formats CSR, MSR, Ellpack-Itpack and diagonal can be used in matrix–vector multiplication. The input format is CSR which can be converted to other formats with calling the subroutines CSRMSR, CSRELL and CSRDIA which were introduced in Appendices A.1.2 to A.1.4. Correspondingly, four subroutines can be called for matrix–vector multiplication as described in Appendices A.2.1 to A.2.4. It should be noted that, one of the matrix–vector multiplication subroutines is allowed to be called; so, calling the remaining three subroutines must be inactive with applying the comment mark at the first of the program executive lines; see lines 41, 46, 51. Here, the subroutine DIAGONAL_MAT_V_PRODUCT is called which multiplies the vector P_OLD by the matrix in diagonal compressed format and results in the vector AP.

Krylov Subspace Methods with Application in Incompressible Fluid Flow Solvers, First Edition.
Iman Farahbakhsh.
© 2020 John Wiley & Sons Ltd. Published 2020 by John Wiley & Sons Ltd.
Companion Website: www.wiley.com/go/Farahbakhs/KrylovSubspaceMethods

```
1   !**************************************************
2   !           Conjugate Gradien Method
3   !**************************************************
4   SUBROUTINE CG (MD,NNZERO,AA_AUX,JA_AUX,IA_AUX,X_OLD,R_OLD,X)
5   USE PARAM, ONLY                     : TOL
6   IMPLICIT NONE
7   INTEGER                             :: K,ITER,TL,MD,NNZERO,IDIAG,IERR
8   REAL(8),DIMENSION(MD)               :: X_OLD,X,P_OLD,P,R_OLD,R,AP
9   INTEGER,DIMENSION(MD+1)             :: IA_AUX,IAR
10  INTEGER,DIMENSION(2*MD-1)           :: IND
11  REAL(8)                             :: SUM_RF,ARF,RFSN,RFS,RF,NORM,S,
        ALPHA,BETA,M,MM,TIME_BEGIN,TIME_END
12  REAL(8),DIMENSION(NNZERO)           :: AA_AUX,AAR
13  INTEGER,DIMENSION(NNZERO)           :: JA_AUX,JAR
14  REAL(8),ALLOCATABLE                 :: SA(:),DIAG(:,:),COEF(:,:)
15  INTEGER,ALLOCATABLE                 :: IJA(:),IOFF(:),JCOEF(:,:)
16  !-------------------------------------------------
17  CALL TOTAL_LENGTH            (MD,IA_AUX,TL)
18  ALLOCATE                     (SA(TL),IJA(TL))
19  CALL CSRMSR                  (TL,MD,NNZERO,AA_AUX,JA_AUX,IA_AUX,SA,IJA)
20  CALL INFDIA                  (MD,NNZERO,JA_AUX,IA_AUX,IND,IDIAG)
21  ALLOCATE                     (COEF(MD,IDIAG),JCOEF(MD,IDIAG))
22  ALLOCATE                     (IOFF(IDIAG),DIAG(MD,IDIAG))
23  CALL CSRELL                  (MD,NNZERO,AA_AUX,JA_AUX,IA_AUX,MD,COEF,
        JCOEF,MD,IDIAG,IERR)
24  CALL CSRDIA                  (MD,NNZERO,IDIAG,10,AA_AUX,JA_AUX,IA_AUX,MD,
        DIAG,IOFF,AAR,JAR,IAR,IND)
25  !-------------------------------------------------
26  P_OLD=R_OLD
27  NORM=1.D0
28  SUM_RF=0.D0
29  ITER=0
30  !CALL CPU_TIME (TIME_BEGIN)
31  DO WHILE (NORM.GT.TOL)
32      ITER=ITER+1
33      !PRINT*,ITER
34      S=0.D0
35      M=0.D0
36      MM=0.D0
37  !-------------------------------------------------
38  ! Stores the matrix in compressed spars row and multiplies by
39  ! vector
40  !-------------------------------------------------
41      !CALL CSR_MAT_V_PRODUCT      (MD,NNZERO,AA_AUX,JA_AUX,IA_AUX,P_OLD,AP)
42  !-------------------------------------------------
43  ! Stores the matrix in modified spars row and multiplies by
44  ! vector
45  !-------------------------------------------------
46      !CALL MSR_MAT_V_PRODUCT (TL,MD,SA,IJA,P_OLD,AP)
47  !-------------------------------------------------
48  ! Stores the matrix in Ellpack/Itpack format and multiplies by
49  ! vector
50  !-------------------------------------------------
51      !CALL ELLPACK_ITPACK_MAT_V_PRODUCT (MD,P_OLD,AP,MD,IDIAG,COEF,JCOEF)
52  !-------------------------------------------------
53  ! Stores the matrix in Diagonal format and multiplies by
54  ! vector
55  !-------------------------------------------------
56      CALL DIAGONAL_MAT_V_PRODUCT (MD,P_OLD,AP,DIAG,MD,IDIAG,IOFF)
57  !-------------------------------------------------
58      DO K=1,MD
59          M=M+R_OLD(K)**2
60          MM=MM+AP(K)*P_OLD(K)
61      END DO
62      ALPHA=M/MM
63      X=X_OLD+ALPHA*P_OLD
64      R=R_OLD-ALPHA*AP
```

```
65  !-----------------------------------------------
66  ! Computing the Euclidean norm
67  !-----------------------------------------------
68      DO K=1,MD
69         S=S+R(K)**2
70      END DO
71      NORM=SQRT(S)/MD
72  !-----------------------------------------------
73  ! Computing the reduction factor
74  !-----------------------------------------------
75      RFS=DOT_PRODUCT(R_OLD,R_OLD)
76      RFSN=DSQRT(RFS)/MD
77      RF=NORM/RFSN
78      SUM_RF=SUM_RF+RF
79  !-----------------------------------------------
80      BETA=S/M
81      P=R+BETA*P_OLD
82      P_OLD=P
83      X_OLD=X
84      R_OLD=R
85      PRINT*,NORM
86      !CALL CPU_TIME (TIME_END)
87      !CALL CPU_TIME_WRITE (TIME_BEGIN,TIME_END,ITER)
88      CALL SUCCESSIVE_SOL (NORM,ITER)
89  END DO
90  ARF=SUM_RF/ITER
91  PRINT*,ARF
92  !-----------------------------------------------
93  END SUBROUTINE CG
```

The constant parameters of the code can be gathered in a module which is called here PARAM. MODULE PARAM contains ITMAX and TOL which denote the maximum number of iterations and convergence tolerance, respectively, and should be manually set by user. This module is called in many parts of the code with the USE command.

```
1  !**************************************************
2  !              Module of Parameters
3  !**************************************************
4  MODULE PARAM
5  !-----------------------------------------------
6  IMPLICIT NONE
7  INTEGER,PARAMETER                    :: ITMAX=1000
8  REAL(8),PARAMETER                    :: TOL=1.D-10
9  !-----------------------------------------------
10 END MODULE PARAM
```

The subroutine TOTAL_LENGTH is called to extract the total length of the real and integer vectors of the MSR format. On entry, MD denotes the matrix dimension and IA is the pointer array of CSR format. This subroutine returns the TL as the total length of MSR format arrays.

```
1  !**************************************************
2  !      Total Length of MSR Format Vectors
3  !**************************************************
4  SUBROUTINE TOTAL_LENGTH (MD,IA,TL)
5  !-----------------------------------------------
```

```
6  IMPLICIT NONE
7  INTEGER                                    :: TL,MD
8  INTEGER,DIMENSION(MD+1)                     :: IA
9  !----------------------------------------------------
10 TL=IA(MD+1)
11 !----------------------------------------------------
12 END SUBROUTINE TOTAL_LENGTH
```

One can compute the computational time of the solution process with calling the intrinsic subroutine CPU_TIME with real arguments TIME_BEGIN and TIME_END in two different lines of the subroutine. SUBROUTINE CPU_TIME_ WRITE writes the total time elapsed since the beginning of the solution in successive iterations.

```
1  !***************************************************
2  !               CPU Time Write
3  !***************************************************
4  SUBROUTINE  CPU_TIME_WRITE (TIME_BEGIN,TIME_END,ITER)
5  !----------------------------------------------------
6  IMPLICIT NONE
7  INTEGER                                    :: ITER,COUNTER
8  CHARACTER*10                                :: EXT
9  CHARACTER*5                                 :: FN1
10 CHARACTER*25                                :: FNAME
11 REAL(8)                                     :: TIME_BEGIN,TIME_END
12 !----------------------------------------------------
13 FN1='CPU_T'
14 COUNTER=0
15 !----------------------------------------------------
16 COUNTER=COUNTER+10
17 WRITE(EXT,'(I7)') ITER
18 FNAME=FN1//EXT//'.DAT'
19 !----------------------------------------------------
20 OPEN(COUNTER,FILE=FNAME,POSITION='REWIND')
21 !WRITE(COUNTER,*)'VARIABLES= "ITER","TIME"'
22 !WRITE(COUNTER,*)'ZONE,F=POINT'
23 WRITE(COUNTER,*) ITER,TIME_END-TIME_BEGIN
24 CLOSE(COUNTER)
25 !----------------------------------------------------
26 !WRITE(*,*)  '========================'
27 !WRITE(*,*)  'PRINTING ON ',FNAME
28 !WRITE(*,*)  '========================'
29 !----------------------------------------------------
30 END SUBROUTINE CPU_TIME_WRITE
```

The residual Euclidean norm and the reduction factor can be saved in each iteration. Subroutine SUCCESSIVE_SOL with two inputs NORM (or RF) and ITER saves the corresponding parameters in the consecutive iterations.

```
1  !***************************************************
2  !            Successive Solution
3  !***************************************************
4  SUBROUTINE SUCCESSIVE_SOL (ERR,ITER)
5  !----------------------------------------------------
```

```
 6 IMPLICIT NONE
 7 INTEGER                                 :: ITER,COUNTER
 8 CHARACTER*10                            :: EXT
 9 CHARACTER*4                             :: FN1
10 CHARACTER*25                            :: FNAME
11 REAL(8)                                 :: ERR
12 !---------------------------------------------------------
13 FN1='VECT'
14 COUNTER=0
15 !---------------------------------------------------------
16 COUNTER=COUNTER+10
17 WRITE(EXT,'(I7)') ITER
18 FNAME=FN1//EXT//'.DAT'
19 !---------------------------------------------------------
20 OPEN(COUNTER,FILE=FNAME,POSITION='REWIND')
21 !WRITE(COUNTER,*)'VARIABLES= "ITER","ERR"'
22 WRITE(COUNTER,*) ITER,ERR
23 CLOSE(COUNTER)
24 !---------------------------------------------------------
25 !WRITE(*,*)  '========================='
26 !WRITE(*,*)  'PRINTING ON ',FNAME
27 !WRITE(*,*)  '========================='
28 !---------------------------------------------------------
29 END SUBROUTINE SUCCESSIVE_SOL
```

It should be noted that, the foregoing subroutines and those mentioned in Appendix A are called in subroutines of Appendices B to D. So, their details will not be presented again.

B.2 Bi-Conjugate Gradient Method

```
 1 !**************************************************
 2 !        Bi-Conjugate Gradiend Method
 3 !**************************************************
 4 SUBROUTINE BCG (MD,NNZERO,AA_AUX,JA_AUX,IA_AUX,X_OLD,R_OLD,X)
 5 USE PARAM, ONLY                         : TOL
 6 IMPLICIT NONE
 7 INTEGER                                 :: K,ITER,MD,NNZERO
 8 REAL(8),DIMENSION(MD)                   :: X_OLD,X,P_OLD,P,R_OLD,R&
 9 ,AP,RS_OLD,PS_OLD,PS,RS,ATPS
10 REAL(8),DIMENSION(NNZERO)               :: AA_AUX
11 INTEGER,DIMENSION(NNZERO)               :: JA_AUX
12 INTEGER,DIMENSION(MD+1)                 :: IA_AUX
13 REAL(8)                                 :: SUM_RF,ARF,RF,RFSN,RFS,NORM,S,
           ALPHA,BETA,M,MM,SN
14 !---------------------------------------------------
15 RS_OLD  = R_OLD
16 P_OLD   = R_OLD
17 PS_OLD  = RS_OLD
18 NORM=1.D0
19 SUM_RF=0.D0
20 ITER=0
21 DO WHILE (NORM.GT.TOL)
22    ITER=ITER+1
```

```
23    !PRINT*,ITER
24    SN=0.D0
25    S=0.D0
26    M=0.D0
27    MM=0.D0
28  !----------------------------------------------------
29    CALL CSR_MAT_V_PRODUCT (MD,NNZERO,AA_AUX,JA_AUX,IA_AUX,P_OLD,AP)
30  !----------------------------------------------------
31    DO K=1,MD
32       M=M+R_OLD(K)*RS_OLD(K)
33       MM=MM+AP(K)*PS_OLD(K)
34    END DO
35    ALPHA=M/MM
36    X=X_OLD+ALPHA*P_OLD
37    R=R_OLD-ALPHA*AP
38  !----------------------------------------------------
39    CALL CSR_TMAT_V_PRODUCT (MD,NNZERO,AA_AUX,JA_AUX,IA_AUX,PS_OLD,ATPS)
40  !----------------------------------------------------
41    RS=RS_OLD-ALPHA*ATPS
42    DO K=1,MD
43       S=S+R(K)*RS(K)
44       SN=SN+R(K)*R(K)
45    END DO
46    NORM=SQRT(SN)/MD
47  !----------------------------------------------------
48  ! Computing the reduction factor
49  !----------------------------------------------------
50    RFS=DOT_PRODUCT(R_OLD,R_OLD)
51    RFSN=DSQRT(RFS)/MD
52    RF=NORM/RFSN
53    SUM_RF=SUM_RF+RF
54  !----------------------------------------------------
55    BETA=S/M
56    P=R+BETA*P_OLD
57    PS=RS+BETA*PS_OLD
58    PS_OLD    =PS
59    P_OLD     =P
60    X_OLD     =X
61    R_OLD     =R
62    RS_OLD    =RS
63    !PRINT*,NORM
64    CALL SUCCESSIVE_SOL (RF,ITER)
65  END DO
66  ARF=SUM_RF/ITER
67  PRINT*,ARF
68  !----------------------------------------------------
69  END SUBROUTINE BCG
```

B.3   Conjugate Gradient Squared Method

```
1   !****************************************************
2   !        Conjugate Gradien Squared Method
3   !****************************************************
4   SUBROUTINE CGS (N,NNZERO,AA_AUX,JA_AUX,IA_AUX,X_OLD,R_OLD,X)
5   !----------------------------------------------------
6   USE PARAM, ONLY             : TOL
7   IMPLICIT NONE
8   INTEGER                     :: K,ITER,N,NNZERO,TL,IDIAG,IERR
9   REAL(8),DIMENSION(NNZERO)   :: AA_AUX,AAR
10  INTEGER,DIMENSION(NNZERO)   :: JA_AUX,JAR
11  INTEGER,DIMENSION(2*N-1)    :: IND
12  INTEGER,DIMENSION(N+1)      :: IA_AUX,IAR
```

```
13 REAL(8),ALLOCATABLE              :: SA(:),COEF(:,:),DIAG(:,:)
14 INTEGER,ALLOCATABLE             :: IJA(:),JCOEF(:,:),IOFF(:)
15 REAL(8),DIMENSION(N)            :: X_OLD,X,P_OLD,P,R_OLD,U,U_OLD,R,
      AP,AUQ,RS,Q
16 REAL(8)                         :: ARF,SUM_RF,RFSN,RF,RFS,NORM,S,
      ALPHA,BETA,M,MM,MN,TIME_BEGIN,TIME_END
17 !-----------------------------------------------
18 CALL TOTAL_LENGTH              (N,IA_AUX,TL)
19 ALLOCATE                       (SA(TL),IJA(TL))
20 CALL CSRMSR                    (TL,N,NNZERO,AA_AUX,JA_AUX,IA_AUX,SA,IJA)
21 CALL INFDIA                    (N,NNZERO,JA_AUX,IA_AUX,IND,IDIAG)
22 ALLOCATE                       (COEF(N,IDIAG),JCOEF(N,IDIAG))
23 ALLOCATE                       (IOFF(IDIAG),DIAG(N,IDIAG))
24 CALL CSRELL                    (N,NNZERO,AA_AUX,JA_AUX,IA_AUX,N,COEF,JCOEF,
      N,IDIAG,IERR)
25 CALL CSRDIA                    (N,NNZERO,IDIAG,10,AA_AUX,JA_AUX,IA_AUX,N,
      DIAG,IOFF,AAR,JAR,IAR,IND)
26 !-----------------------------------------------
27 RS=R_OLD
28 P_OLD=R_OLD
29 U_OLD=R_OLD
30 NORM=1.D0
31 SUM_RF=0.D0
32 !CALL CPU_TIME (TIME_BEGIN)
33 ITER=0
34 DO WHILE (NORM.GT.TOL)
35    ITER=ITER+1
36    !PRINT*,ITER
37    S=0.D0
38    M=0.D0
39    MM=0.D0
40    MN=0.D0
41 !-----------------------------------------------
42 ! Stores the matrix in compressed spars row and multiplies by
43 ! vector
44 !-----------------------------------------------
45    !CALL CSR_MAT_V_PRODUCT       (N,NNZERO,AA_AUX,JA_AUX,IA_AUX,P_OLD,AP)
46 !-----------------------------------------------
47 ! Stores the matrix in modified spars row and multiplies by
48 ! vector
49 !-----------------------------------------------
50    !CALL MSR_MAT_V_PRODUCT (TL,N,SA,IJA,P_OLD,AP)
51 !-----------------------------------------------
52 ! Stores the matrix in Ellpack/Itpack format and multiplies by
53 ! vector
54 !-----------------------------------------------
55    !CALL ELLPACK_ITPACK_MAT_V_PRODUCT (N,P_OLD,AP,N,IDIAG,COEF,JCOEF)
56 !-----------------------------------------------
57 ! Stores the matrix in Diagonal format and multiplies by
58 ! vector
59 !-----------------------------------------------
60    CALL DIAGONAL_MAT_V_PRODUCT (N,P_OLD,AP,DIAG,N,IDIAG,IOFF)
61 !-----------------------------------------------
62    DO K=1,N
63       M=M+R_OLD(K)*RS(K)
64       MM=MM+AP(K)*RS(K)
65    END DO
66    ALPHA=M/MM
67    Q=U_OLD-ALPHA*AP
68    X=X_OLD+ALPHA*(U_OLD+Q)
69 !-----------------------------------------------
70 ! Stores the matrix in compressed spars row and multiplies by
71 ! vector
72 !-----------------------------------------------
73    !CALL CSR_MAT_V_PRODUCT       (N,NNZERO,AA_AUX,JA_AUX,IA_AUX,U_OLD+Q,
      AUQ)
74 !-----------------------------------------------
```

```
75  ! Stores the matrix in modified spars row and multiplies by
76  ! vector
77  !---------------------------------------------------
78     !CALL MSR_MAT_V_PRODUCT (TL,N,SA,IJA,U_OLD+Q,AUQ)
79  !---------------------------------------------------
80  ! Stores the matrix in Ellpack/Itpack format and multiplies by
81  ! vector
82  !---------------------------------------------------
83     !CALL ELLPACK_ITPACK_MAT_V_PRODUCT (N,U_OLD+Q,AUQ,N,IDIAG,COEF,JCOEF
       )
84  !---------------------------------------------------
85  ! Stores the matrix in Diagonal format and multiplies by
86  ! vector
87  !---------------------------------------------------
88     CALL DIAGONAL_MAT_V_PRODUCT (N,U_OLD+Q,AUQ,DIAG,N,IDIAG,IOFF)
89  !---------------------------------------------------
90     R=R_OLD-ALPHA*AUQ
91     DO K=1,N
92        S=S+R(K)**2
93     END DO
94     NORM=SQRT(S)/N
95  !---------------------------------------------------
96  ! Computing the reduction factor
97  !---------------------------------------------------
98     RFS=DOT_PRODUCT(R_OLD,R_OLD)
99     RFSN=DSQRT(RFS)/N
100    RF=NORM/RFSN
101    SUM_RF=SUM_RF+RF
102 !---------------------------------------------------
103    DO K=1,N
104       MN=MN+R(K)*RS(K)
105    END DO
106    BETA=MN/M
107    U=R+BETA*Q
108    P=U+BETA*(Q+BETA*P_OLD)
109    P_OLD=P
110    X_OLD=X
111    R_OLD=R
112    U_OLD=U
113    PRINT*,NORM
114    !CALL CPU_TIME (TIME_END)
115    !CALL CPU_TIME_WRITE (TIME_BEGIN,TIME_END,ITER)
116    CALL SUCCESSIVE_SOL (RF,ITER)
117 END DO
118 ARF=SUM_RF/ITER
119 PRINT*,ARF
120 !---------------------------------------------------
121 END SUBROUTINE CGS
```

B.4   Bi-Conjugate Gradient Stabilized Method

```
1  !****************************************************
2  !     Bi-Conjugate Gradien Stabilized Method
3  !****************************************************
4  SUBROUTINE BCGSTAB (N,NNZERO,AA_AUX,JA_AUX,IA_AUX,X_OLD,R_OLD,X)
5  !---------------------------------------------------
6  USE PARAM, ONLY            : TOL
7  IMPLICIT NONE
8  INTEGER                    :: K,ITER,N,NNZERO,TL,IDIAG,IERR
9  REAL(8),DIMENSION(NNZERO)  :: AA_AUX,AAR
10 INTEGER,DIMENSION(NNZERO)  :: JA_AUX,JAR
11 INTEGER,DIMENSION(2*N-1)   :: IND
12 INTEGER,DIMENSION(N+1)     :: IA_AUX,IAR
```

```
13  REAL(8),ALLOCATABLE                :: SA(:),COEF(:,:),DIAG(:,:)
14  INTEGER,ALLOCATABLE                :: IJA(:),JCOEF(:,:),IOFF(:)
15  REAL(8),DIMENSION(N)               :: X_OLD,X,P_OLD,P,R_OLD,R,AP,AQQ,RS
        ,QQ
16  REAL(8)                            :: SUM_RF,ARF,RF,RFSN,RFS,NORM,S,
        ALPHA,BETA,M,MS,MMS,MM,MN,TIME_BEGIN,TIME_END,OM
17  !-------------------------------------------------------
18  CALL TOTAL_LENGTH                  (N,IA_AUX,TL)
19  ALLOCATE                           (SA(TL),IJA(TL))
20  CALL CSRMSR                        (TL,N,NNZERO,AA_AUX,JA_AUX,IA_AUX,SA,IJA)
21  CALL INFDIA                        (N,NNZERO,JA_AUX,IA_AUX,IND,IDIAG)
22  ALLOCATE                           (COEF(N,IDIAG),JCOEF(N,IDIAG))
23  ALLOCATE                           (IOFF(IDIAG),DIAG(N,IDIAG))
24  CALL CSRELL                        (N,NNZERO,AA_AUX,JA_AUX,IA_AUX,N,COEF,JCOEF,
        N,IDIAG,IERR)
25  CALL CSRDIA                        (N,NNZERO,IDIAG,10,AA_AUX,JA_AUX,IA_AUX,N,
        DIAG,IOFF,AAR,JAR,IAR,IND)
26  !-------------------------------------------------------
27  RS=R_OLD
28  P_OLD=R_OLD
29  NORM=1.D0
30  SUM_RF=0.D0
31  ITER=0
32  !CALL CPU_TIME (TIME_BEGIN)
33  DO WHILE (NORM.GT.TOL)
34     ITER=ITER+1
35     !PRINT*,ITER
36     S=0.D0
37     M=0.D0
38     MS=0.D0
39     MMS=0.D0
40     MM=0.D0
41     MN=0.D0
42  !-------------------------------------------------------
43  ! Stores the matrix in compressed spars row and multiplies by
44  ! vector
45  !-------------------------------------------------------
46     !CALL CSR_MAT_V_PRODUCT      (N,NNZERO,AA_AUX,JA_AUX,IA_AUX,P_OLD,AP)
47  !-------------------------------------------------------
48  ! Stores the matrix in modified spars row and multiplies by
49  ! vector
50  !-------------------------------------------------------
51     !CALL MSR_MAT_V_PRODUCT (TL,N,SA,IJA,P_OLD,AP)
52  !-------------------------------------------------------
53  ! Stores the matrix in Ellpack/Itpack format and multiplies by
54  ! vector
55  !-------------------------------------------------------
56     !CALL ELLPACK_ITPACK_MAT_V_PRODUCT (N,P_OLD,AP,N,IDIAG,COEF,JCOEF)
57  !-------------------------------------------------------
58  ! Stores the matrix in Diagonal format and multiplies by
59  ! vector
60  !-------------------------------------------------------
61     CALL DIAGONAL_MAT_V_PRODUCT (N,P_OLD,AP,DIAG,N,IDIAG,IOFF)
62  !-------------------------------------------------------
63     DO K=1,N
64        M=M+R_OLD(K)*RS(K)
65        MM=MM+AP(K)*RS(K)
66     END DO
67     ALPHA=M/MM
68     QQ=R_OLD-ALPHA*AP
69  !-------------------------------------------------------
70  ! Stores the matrix in compressed spars row and multiplies by
71  ! vector
72  !-------------------------------------------------------
73     !CALL CSR_MAT_V_PRODUCT      (N,NNZERO,AA_AUX,JA_AUX,IA_AUX,QQ,AQQ)
74  !-------------------------------------------------------
75  ! Stores the matrix in modified spars row and multiplies by
```

```fortran
76   ! vector
77   !-------------------------------------------------------
78       !CALL MSR_MAT_V_PRODUCT (TL,N,SA,IJA,QQ,AQQ)
79   !-------------------------------------------------------
80   ! Stores the matrix in Ellpack/Itpack format and multiplies by
81   ! vector
82   !-------------------------------------------------------
83       !CALL ELLPACK_ITPACK_MAT_V_PRODUCT (N,QQ,AQQ,N,IDIAG,COEF,JCOEF)
84   !-------------------------------------------------------
85   ! Stores the matrix in Diagonal format and multiplies by
86   ! vector
87   !-------------------------------------------------------
88       CALL DIAGONAL_MAT_V_PRODUCT (N,QQ,AQQ,DIAG,N,IDIAG,IOFF)
89   !-------------------------------------------------------
90       DO K=1,N
91          MS=MS+AQQ(K)*QQ(K)
92          MMS=MMS+AQQ(K)*AQQ(K)
93       END DO
94       OM=MS/MMS
95       X=X_OLD+ALPHA*P_OLD+OM*QQ
96       R=QQ-OM*AQQ
97       DO K=1,N
98          S=S+R(K)**2
99       END DO
100      NORM=SQRT(S)/N
101  !-------------------------------------------------------
102  ! Computing the reduction factor
103  !-------------------------------------------------------
104      RFS=DOT_PRODUCT(R_OLD,R_OLD)
105      RFSN=DSQRT(RFS)/N
106      RF=NORM/RFSN
107      SUM_RF=SUM_RF+RF
108  !-------------------------------------------------------
109      DO K=1,N
110         MN=MN+R(K)*RS(K)
111      END DO
112      BETA=(MN/M)*(ALPHA/OM)
113      P=R+BETA*(P_OLD-OM*AP)
114      P_OLD=P
115      X_OLD=X
116      R_OLD=R
117      !PRINT*,NORM
118      !CALL CPU_TIME (TIME_END)
119      !CALL CPU_TIME_WRITE (TIME_BEGIN,TIME_END,ITER)
120      CALL SUCCESSIVE_SOL (RF,ITER)
121  END DO
122  ARF=SUM_RF/ITER
123  PRINT*,ARF
124  !-------------------------------------------------------
125  END SUBROUTINE BCGSTAB
```

B.5 Conjugate Residual Method

```fortran
1    !****************************************************
2    !        Conjugate Residual Subroutine
3    !****************************************************
4    SUBROUTINE CR (MD,NNZERO,AA_AUX,JA_AUX,IA_AUX,X_OLD,R_OLD,X)
5    !-------------------------------------------------------
6    USE PARAM, ONLY                    : TOL
7    IMPLICIT NONE
8    INTEGER                            :: ITER,TL,MD,NNZERO,IDIAG,IERR
9    REAL(8),DIMENSION(MD)              :: X_OLD,X,P_OLD,P,R_OLD,R,AP,AP_OLD
          ,AR_OLD,AR
```

```
10 INTEGER,DIMENSION(MD+1)                :: IA_AUX,IAR
11 INTEGER,DIMENSION(2*MD-1)              :: IND
12 REAL(8)                                :: SUM_RF,ARF,RF,RFSN,RFS,NORM,S,
      ALPHA,BETA,TIME_BEGIN,TIME_END
13 REAL(8),DIMENSION(NNZERO)              :: AA_AUX,AAR
14 INTEGER,DIMENSION(NNZERO)              :: JA_AUX,JAR
15 REAL(8),ALLOCATABLE                    :: SA(:),DIAG(:,:),COEF(:,:)
16 INTEGER,ALLOCATABLE                    :: IJA(:),IOFF(:),JCOEF(:,:)
17 !------------------------------------------------------
18 CALL TOTAL_LENGTH          (MD,IA_AUX,TL)
19 ALLOCATE                   (SA(TL),IJA(TL))
20 CALL CSRMSR                (TL,MD,NNZERO,AA_AUX,JA_AUX,IA_AUX,SA,IJA)
21 CALL INFDIA                (MD,NNZERO,JA_AUX,IA_AUX,IND,IDIAG)
22 ALLOCATE                   (COEF(MD,IDIAG),JCOEF(MD,IDIAG))
23 ALLOCATE                   (IOFF(IDIAG),DIAG(MD,IDIAG))
24 CALL CSRELL                (MD,NNZERO,AA_AUX,JA_AUX,IA_AUX,MD,COEF,
      JCOEF,MD,IDIAG,IERR)
25 CALL CSRDIA                (MD,NNZERO,IDIAG,10,AA_AUX,JA_AUX,IA_AUX,MD,
      DIAG,IOFF,AAR,JAR,IAR,IND)
26 !------------------------------------------------------
27 P_OLD=R_OLD
28 CALL MSR_MAT_V_PRODUCT (TL,MD,SA,IJA,P_OLD,AP_OLD)
29 NORM=1.D0
30 SUM_RF=0.D0
31 ITER=0
32 !CALL CPU_TIME (TIME_BEGIN)
33 DO WHILE (NORM.GT.TOL)
34    ITER=ITER+1
35    !PRINT*,ITER
36 !------------------------------------------------------
37 ! Stores the matrix in compressed spars row and multiplies by
38 ! vector
39 !------------------------------------------------------
40    !CALL CSR_MAT_V_PRODUCT    (MD,NNZERO,AA_AUX,JA_AUX,IA_AUX,R_OLD,
      AR_OLD)
41 !------------------------------------------------------
42 ! Stores the matrix in modified spars row and multiplies by
43 ! vector
44 !------------------------------------------------------
45    !CALL MSR_MAT_V_PRODUCT (TL,MD,SA,IJA,R_OLD,AR_OLD)
46 !------------------------------------------------------
47 ! Stores the matrix in Ellpack/Itpack format and multiplies by
48 ! vector
49 !------------------------------------------------------
50    !CALL ELLPACK_ITPACK_MAT_V_PRODUCT (MD,R_OLD,AR_OLD,MD,IDIAG,COEF,
      JCOEF)
51 !------------------------------------------------------
52 ! Stores the matrix in Diagonal format and multiplies by
53 ! vector
54 !------------------------------------------------------
55    CALL DIAGONAL_MAT_V_PRODUCT (MD,R_OLD,AR_OLD,DIAG,MD,IDIAG,IOFF)
56 !------------------------------------------------------
57    ALPHA=DOT_PRODUCT(R_OLD,AR_OLD)/DOT_PRODUCT(AP_OLD,AP_OLD)
58    X=X_OLD+ALPHA*P_OLD
59    R=R_OLD-ALPHA*AP_OLD
60 !------------------------------------------------------
61 ! Stores the matrix in compressed spars row and multiplies by
62 ! vector
63 !------------------------------------------------------
64    !CALL CSR_MAT_V_PRODUCT    (MD,NNZERO,AA_AUX,JA_AUX,IA_AUX,R,AR)
65 !------------------------------------------------------
66 ! Stores the matrix in modified spars row and multiplies by
67 ! vector
68 !------------------------------------------------------
69    !CALL MSR_MAT_V_PRODUCT (TL,MD,SA,IJA,R,AR)
70 !------------------------------------------------------
71 ! Stores the matrix in Ellpack/Itpack format and multiplies by
```

```
72  ! vector
73  !--------------------------------------------------------
74      !CALL ELLPACK_ITPACK_MAT_V_PRODUCT (MD,R,AR,MD,IDIAG,COEF,JCOEF)
75  !--------------------------------------------------------
76  ! Stores the matrix in Diagonal format and multiplies by
77  ! vector
78  !--------------------------------------------------------
79      CALL DIAGONAL_MAT_V_PRODUCT (MD,R,AR,DIAG,MD,IDIAG,IOFF)
80  !--------------------------------------------------------
81      BETA=DOT_PRODUCT(R,AR)/DOT_PRODUCT(R_OLD,AR_OLD)
82      P=R+BETA*P_OLD
83      AP=AR+BETA*AP_OLD
84      S=DOT_PRODUCT(R,R)
85      NORM=DSQRT(S)/MD
86  !--------------------------------------------------------
87  ! Computing the reduction factor
88  !--------------------------------------------------------
89      RFS=DOT_PRODUCT(R_OLD,R_OLD)
90      RFSN=DSQRT(RFS)/MD
91      RF=NORM/RFSN
92      SUM_RF=SUM_RF+RF
93  !--------------------------------------------------------
94      P_OLD=P
95      X_OLD=X
96      R_OLD=R
97      AP_OLD=AP
98      PRINT*,NORM
99      !CALL CPU_TIME (TIME_END)
100     !CALL CPU_TIME_WRITE (TIME_BEGIN,TIME_END,ITER)
101     CALL SUCCESSIVE_SOL (RF,ITER)
102 END DO
103 ARF=SUM_RF/ITER
104 PRINT*,ARF
105 !--------------------------------------------------------
106 END SUBROUTINE CR
```

B.6   GMRES* Method

In this appendix, subroutine GMRES* is introduced. The distinctive aspect of this subroutine is the use of other Krylov subspace methods as a preconditioner; see Section 3.6. The inner iterations of the CG, CGS, BiCGSTAB, CR, and preconditioned versions of the first three ones, can be called by user via uncommenting a corresponding line, i.e., lines 30 to 36. See Appendix D for details of the inner iteration subroutines of the Krylov subspace methods.

```
1  !****************************************************
2  !                GMRES* Method
3  !****************************************************
4  SUBROUTINE GMRES_STAR (MD,NNZERO,AA_AUX,JA_AUX,IA_AUX,XX_OLD,R_OLD,X)
5  !--------------------------------------------------------
6  USE PARAM, ONLY                    : TOL,ITMAX
7  IMPLICIT NONE
8  INTEGER                            :: I,K,MD,NNZERO,IERR,TL,IDIAG
9  REAL(8)                            :: NORMC,NORM,ALPHA,CR,S,SR
10 REAL(8),ALLOCATABLE                :: SA(:),DIAG(:,:),COEF(:,:)
11 INTEGER,ALLOCATABLE                :: IJA(:),IOFF(:),JCOEF(:,:)
12 REAL(8),DIMENSION(NNZERO)          :: AA_AUX,AAR
13 INTEGER,DIMENSION(NNZERO)          :: JA_AUX,JAR
14 INTEGER,DIMENSION(MD+1)            :: IA_AUX,IAR
```

```
15  INTEGER,DIMENSION(2*MD-1)              :: IND
16  REAL(8),DIMENSION(MD)                  :: ZM,X_OLD,XX_OLD,R_OLD,X,R,C
17  REAL(8),DIMENSION(ITMAX,MD)            :: CC,U
18  !-------------------------------------------------------
19  CALL TOTAL_LENGTH            (MD,IA_AUX,TL)
20  ALLOCATE                     (SA(TL),IJA(TL))
21  CALL CSRMSR                  (TL,MD,NNZERO,AA_AUX,JA_AUX,IA_AUX,SA,IJA)
22  CALL INFDIA                  (MD,NNZERO,JA_AUX,IA_AUX,IND,IDIAG)
23  ALLOCATE                     (COEF(MD,IDIAG),JCOEF(MD,IDIAG))
24  ALLOCATE                     (IOFF(IDIAG),DIAG(MD,IDIAG))
25  CALL CSRELL                  (MD,NNZERO,AA_AUX,JA_AUX,IA_AUX,MD,COEF,
        JCOEF,MD,IDIAG,IERR)
26  CALL CSRDIA                  (MD,NNZERO,IDIAG,10,AA_AUX,JA_AUX,IA_AUX,MD,
        DIAG,IOFF,AAR,JAR,IAR,IND)
27  !-------------------------------------------------------
28  DO I=1,ITMAX
29  !-------------------------------------------------------
30      !CALL CG_INNER               (MD,NNZERO,20,AA_AUX,JA_AUX,IA_AUX,XX_OLD,
        R_OLD,ZM)
31      !CALL ILU0_PCG_INNER         (MD,NNZERO,20,AA_AUX,JA_AUX,IA_AUX,XX_OLD,
        R_OLD,ZM)
32      !CALL BCGSTAB_INNER          (MD,NNZERO,20,AA_AUX,JA_AUX,IA_AUX,XX_OLD,
        R_OLD,ZM)
33      !CALL ILU0_PBCGSTAB_INNER    (MD,NNZERO,20,AA_AUX,JA_AUX,IA_AUX,XX_OLD,
        R_OLD,ZM)
34      !CALL CGS_INNER              (MD,NNZERO,20,AA_AUX,JA_AUX,IA_AUX,XX_OLD,
        R_OLD,ZM)
35      !CALL ILU0_PCGS_INNER        (MD,NNZERO,20,AA_AUX,JA_AUX,IA_AUX,XX_OLD,
        R_OLD,ZM)
36      CALL CR_INNER               (MD,NNZERO,20,AA_AUX,JA_AUX,IA_AUX,XX_OLD,
        R_OLD,ZM)
37  !-------------------------------------------------------
38  ! Stores the matrix in compressed spars row and multiplies by
39  ! vector
40  !-------------------------------------------------------
41  !CALL CSR_MAT_V_PRODUCT     (MD,NNZERO,AA_AUX,JA_AUX,IA_AUX,ZM,C)
42  !-------------------------------------------------------
43  ! Stores the matrix in modified spars row and multiplies by
44  ! vector
45  !-------------------------------------------------------
46  !CALL MSR_MAT_V_PRODUCT (TL,MD,SA,IJA,ZM,C)
47  !-------------------------------------------------------
48  ! Stores the matrix in Ellpack/Itpack format and multiplies by
49  ! vector
50  !-------------------------------------------------------
51  !CALL ELLPACK_ITPACK_MAT_V_PRODUCT (MD,ZM,C,MD,IDIAG,COEF,JCOEF)
52  !-------------------------------------------------------
53  ! Stores the matrix in Diagonal format and multiplies by
54  ! vector
55  !-------------------------------------------------------
56      CALL DIAGONAL_MAT_V_PRODUCT (MD,ZM,C,DIAG,MD,IDIAG,IOFF)
57  !-------------------------------------------------------
58      DO K=1,I-1
59          ALPHA=DOT_PRODUCT(CC(K,:),C)
60          C=C-ALPHA*CC(K,:)
61          ZM=ZM-ALPHA*U(K,:)
62      END DO
63      S=DOT_PRODUCT(C,C)
64      NORMC=DSQRT(S)
65      CC(I,:)=C/NORMC
66      U(I,:)=ZM/NORMC
67      CR=DOT_PRODUCT(CC(I,:),R_OLD)
68      X=X_OLD+CR*U(I,:)
69      R=R_OLD-CR*CC(I,:)
70      SR=DOT_PRODUCT(R,R)
71      NORM=DSQRT(SR)/MD
72      R_OLD=R
```

```
73    X_OLD=X
74    !PRINT*,NORM
75    !CALL SUCCESSIVE_SOL (NORM,I)
76    IF (NORM.LT.TOL) THEN
77       EXIT
78    ELSE IF (I.EQ.ITMAX.AND.NORM.GT.TOL) THEN
79       PRINT*, 'convergency is not met by this number of iteration'
80    END IF
81 END DO
82 !--------------------------------------------------
83 END SUBROUTINE GMRES_STAR
```

Appendix C

ILU(0) Preconditioning

C.1 ILU(0)-Preconditioned Conjugate Gradient Method

The incomplete Cholesky factorization with zero fill or ILU(0), see Section 4.2.1, is applied as the preconditioner for the CG, CGS and BiCGSTAB methods in the current and following sections, respectively. The argument of the preconditioned methods subroutines are similar to the subroutines mentioned in Appendix B. We introduce subroutines ILU0 and LUSOL which are called in subroutines ILU0_PCG, ILU0_PCGS and ILU0_PBCGSTAB. Because of the similarity of the subroutines argument list and nested calls, we only comment on this section and refrain from repeating it in Appendices C.2 and C.3.

```
1  !***************************************************
2  !ILU(0)-Preconditioned Conjugate Gradient Method
3  !***************************************************
4  SUBROUTINE ILU0_PCG (N,NNZERO,AA_AUX,JA_AUX,IA_AUX,X_OLD,R_OLD,X)
5  !-------------------------------------------------
6  USE PARAM, ONLY                  : TOL
7  IMPLICIT NONE
8  INTEGER                     :: K,ITER,TL,N,NNZERO,ICODE,IDIAG,IERR
9  REAL(8),DIMENSION(N)        :: X_OLD,X,P_OLD,P,R_OLD,R,AP,Z_OLD,Z
10 INTEGER,DIMENSION(N+1)      :: IA_AUX,IAR
11 INTEGER,DIMENSION(2*N-1)    :: IND
12 REAL(8)                     :: NORM,S,ALPHA,BETA,M,MM
13 REAL(8),DIMENSION(NNZERO)   :: AA_AUX,AAR,LUVAL
14 INTEGER,DIMENSION(NNZERO)   :: JA_AUX,JAR
15 REAL(8),ALLOCATABLE         :: SA(:),COEF(:,:),DIAG(:,:)
16 INTEGER,ALLOCATABLE         :: IJA(:),JCOEF(:,:),IOFF(:)
17 INTEGER,DIMENSION(N)        :: UPTR,IW
18 REAL(8)                     :: SUM_RF,ARF,RF,RFSN,RFS,SN,
      TIME_BEGIN,TIME_END
19 !-------------------------------------------------
20 CALL TOTAL_LENGTH  (N,IA_AUX,TL)
21 ALLOCATE           (SA(TL),IJA(TL))
22 CALL CSRMSR        (TL,N,NNZERO,AA_AUX,JA_AUX,IA_AUX,SA,IJA)
23 CALL INFDIA        (N,NNZERO,JA_AUX,IA_AUX,IND,IDIAG)
24 ALLOCATE           (COEF(N,IDIAG),JCOEF(N,IDIAG))
25 ALLOCATE           (IOFF(IDIAG),DIAG(N,IDIAG))
26 CALL CSRELL        (N,NNZERO,AA_AUX,JA_AUX,IA_AUX,N,COEF,JCOEF,N,
      IDIAG,IERR)
```

Krylov Subspace Methods with Application in Incompressible Fluid Flow Solvers, First Edition. Iman Farahbakhsh.
© 2020 John Wiley & Sons Ltd. Published 2020 by John Wiley & Sons Ltd.
Companion Website: www.wiley.com/go/Farahbakhs/KrylovSubspaceMethods

```
27  CALL CSRDIA      (N,NNZERO,IDIAG,10,AA_AUX,JA_AUX,IA_AUX,N,DIAG,
        IOFF,AAR,JAR,IAR,IND)
28  !-------------------------------------------------
29  CALL ILU0    (N,NNZERO,AA_AUX,JA_AUX,IA_AUX,LUVAL,UPTR,IW,ICODE)
30  CALL LUSOL   (N,NNZERO,R_OLD,Z_OLD,LUVAL,JA_AUX,IA_AUX,UPTR)
31  !-------------------------------------------------
32  P_OLD=Z_OLD
33  NORM=1.D0
34  SUM_RF=0.D0
35  ITER=0
36  !CALL CPU_TIME (TIME_BEGIN)
37  DO WHILE (NORM.GT.TOL)
38      ITER=ITER+1
39      !PRINT*,ITER
40      SN=0.D0
41      S=0.D0
42      M=0.D0
43      MM=0.D0
44  !-------------------------------------------------------------
45  ! Stores the matrix in compressed spars row and multiplies by
46  ! vector
47  !-------------------------------------------------------
48  !CALL CSR_MAT_V_PRODUCT    (N,NNZERO,AA_AUX,JA_AUX,IA_AUX,P_OLD,AP
        )
49  !-------------------------------------------------------
50  ! Stores the matrix in modified spars row and multiplies by
51  ! vector
52  !-------------------------------------------------
53  !CALL MSR_MAT_V_PRODUCT (TL,N,SA,IJA,P_OLD,AP)
54  !-------------------------------------------------------
55  ! Stores the matrix in Ellpack/Itpack format and multiplies by
56  ! vector
57  !-------------------------------------------------------
58  !CALL ELLPACK_ITPACK_MAT_V_PRODUCT (N,P_OLD,AP,N,IDIAG,COEF,JCOEF)
59  !-------------------------------------------------------
60  ! Stores the matrix in Diagonal format and multiplies by
61  ! vector
62  !-------------------------------------------------------
63      CALL DIAGONAL_MAT_V_PRODUCT (N,P_OLD,AP,DIAG,N,IDIAG,IOFF)
64  !-------------------------------------------------------
65      DO K=1,N
66          M=M+R_OLD(K)*Z_OLD(K)
67          MM=MM+AP(K)*P_OLD(K)
68      END DO
69      ALPHA=M/MM
70      X=X_OLD+ALPHA*P_OLD
71      R=R_OLD-ALPHA*AP
72  !-------------------------------------------------------
73      CALL LUSOL   (N,NNZERO,R,Z,LUVAL,JA_AUX,IA_AUX,UPTR)
74  !-------------------------------------------------
75      DO K=1,N
76          S=S+R(K)*Z(K)
77      END DO
78      BETA=S/M
79      P=Z+BETA*P_OLD
80      DO K=1,N
81          SN=SN+R(K)**2
82      END DO
83      NORM=SQRT(SN)/N
84  !-------------------------------------------------------
85  ! Computing the reduction factor
86  !-------------------------------------------------------
87      RFS=DOT_PRODUCT(R_OLD,R_OLD)
```

```
88      RFSN=DSQRT(RFS)/N
89      RF=NORM/RFSN
90      SUM_RF=SUM_RF+RF
91    !-------------------------------------------------
92      P_OLD=P
93      X_OLD=X
94      R_OLD=R
95      Z_OLD=Z
96    !PRINT*,NORM
97    !CALL CPU_TIME (TIME_END)
98    !CALL CPU_TIME_WRITE (TIME_BEGIN,TIME_END,ITER)
99    !CALL SUCCESSIVE_SOL (NORM,ITER)
100   END DO
101   !ARF=SUM_RF/ITER
102   !PRINT*,ARF
103   !-------------------------------------------------
104   END SUBROUTINE ILU0_PCG
```

This subroutine computes the *L* and *U* factors of the ILU(0) factorization of a general sparse matrix *A* stored in CSR format. Since *L* is unit triangular, the *L* and *U* factors can be stored as a single matrix that occupies the same storage as *A*. The JA and IA arrays are not needed for the *LU* matrix since the pattern of the *LU* matrix is identical to that of *A*.

On entry, N and NNZERO denote the dimension of the matrix and the number of non-zero elements, respectively. The matrix in CSR format is presented with three vectors A, JA and IA. IW is an integer work array of length N. On return, *L* and *U* matrices are stored together. LUVAL, JA and IA is the combined CSR data structure for the *LU* factors. UPTR is the pointer to the diagonal elements in the CSR data structure that is LUVAL, JA and IA. ICODE is an integer indicating error code on return. The zero value of ICODE shows the normal return and ICODE=K denotes that return is encountered a zero pivot at step K.

```
1    !***********************************************************
2    !   Incomplete LU Factorization with 0 Level of Fill in
3    !***********************************************************
4    SUBROUTINE ILU0 (N,NNZERO,A,JA,IA,LUVAL,UPTR,IW,ICODE)
5    !----------------------------------------------------
6    IMPLICIT NONE
7    INTEGER                       :: N,I,J,K,J1,J2,ICODE,JROW,NNZERO,JJ,
                                        JW
8    INTEGER,DIMENSION(NNZERO)     :: JA
9    INTEGER,DIMENSION(N+1)        :: IA
10   INTEGER,DIMENSION(N)          :: IW,UPTR
11   REAL(8),DIMENSION(NNZERO)     :: A
12   REAL(8),DIMENSION(NNZERO)     :: LUVAL
13   REAL(8)                       :: T
14   !----------------------------------------------------
15   DO I=1,IA(N+1)-1
16      LUVAL(I)=A(I)
17   END DO
18   DO I=1,N
19      IW(I)=0
20   END DO
21   !----------------------------------------------------
22   DO K=1,N
23      J1=IA(K)
24      J2=IA(K+1)-1
```

```
25        DO J=J1,J2
26           IW(JA(J))=J
27        END DO
28           J=J1
29 150    JROW=JA(J)
30 !----------------------------------------------
31        IF (JROW.GE.K) GO TO 200
32 !----------------------------------------------
33        T=LUVAL(J)*LUVAL(UPTR(JROW))
34        LUVAL(J)=T
35 !----------------------------------------------
36        DO JJ=UPTR(JROW)+1,IA(JROW+1)-1
37           JW=IW(JA(JJ))
38           IF (JW.NE.0) LUVAL(JW)=LUVAL(JW)-T*LUVAL(JJ)
39        END DO
40        J=J+1
41        IF (J.LE.J2) GO TO 150
42 !----------------------------------------------
43 200    UPTR(K)=J
44        IF (JROW.NE.K.OR.LUVAL(J).EQ.0.D0) GO TO 600
45        LUVAL(J)=1.D0/LUVAL(J)
46 !----------------------------------------------
47        DO I=J1,J2
48           IW(JA(I))=0
49        END DO
50 END DO
51 !----------------------------------------------
52 ICODE=0
53 RETURN
54 !----------------------------------------------
55 600    ICODE=K
56 RETURN
57 !----------------------------------------------
58 END SUBROUTINE ILU0
```

Subroutine LUSOL solves (LU) SOL=RHS in a forward and a backward process. On entry, N and NNZERO denotes the matrix dimension and number of non-zero elements, respectively. Vector RHS is the right-hand side vector that is unchanged on return. The values of the *LU* matrix are stored in LUVAL. *L* and *U* are stored together in CSR format. The diagonal elements of *U* are inverted. In each row, the *L* values are followed by the diagonal element (inverted) and then the other *U* values. LUCOL denotes the column indices of corresponding elements in LUVAL. LUPTR contains the pointers to the beginning of each row in the *LU* matrix and UPTR denotes the pointer to the diagonal elements in LUVAL and LUCOL. On return, there is just vector SOL that is the solution of system (LU) SOL=RHS.

```
1  !*************************************************************
2  !                     LU Solution
3  !*************************************************************
4  SUBROUTINE LUSOL (N,NNZERO,RHS,SOL,LUVAL,LUCOL,LUPTR,UPTR)
5  !----------------------------------------------
6  IMPLICIT NONE
7  INTEGER                     :: I,K,N,NNZERO
8  REAL(8),DIMENSION(NNZERO)   :: LUVAL
9  REAL(8),DIMENSION(N)        :: SOL,RHS
10 INTEGER,DIMENSION(N)        :: UPTR
11 INTEGER,DIMENSION(N+1)      :: LUPTR
12 INTEGER,DIMENSION(NNZERO)   :: LUCOL
13 !----------------------------------------------
```

```
14  DO I=1,N
15     SOL(I)=RHS(I)
16     DO K=LUPTR(I),UPTR(I)-1
17        SOL(I)=SOL(I)-LUVAL(K)*SOL(LUCOL(K))
18     END DO
19  END DO
20  !---------------------------------------------
21  DO I=N,1,-1
22     DO K=UPTR(I)+1,LUPTR(I+1)-1
23        SOL(I)=SOL(I)-LUVAL(K)*SOL(LUCOL(K))
24     END DO
25     SOL(I)=LUVAL(UPTR(I))*SOL(I)
26  END DO
27  !---------------------------------------------
28  END SUBROUTINE LUSOL
```

C.2   ILU(0)-Preconditioned Conjugate Gradient Squared Method

```
1   !******************************************************
2   !ILU(0)-Preconditioned Conjugate Gradient Squared Method
3   !******************************************************
4   SUBROUTINE ILU0_PCGS (N,NNZERO,AA_AUX,JA_AUX,IA_AUX,X_OLD,R_OLD,X)
5   !---------------------------------------------
6   USE PARAM, ONLY           : TOL
7   IMPLICIT NONE
8   INTEGER                   :: ITER,N,NNZERO,TL,ICODE,IDIAG
        ,IERR
9   REAL(8),DIMENSION(NNZERO)    :: AA_AUX,AAR,LUVAL
10  INTEGER,DIMENSION(NNZERO)    :: JA_AUX,JAR
11  INTEGER,DIMENSION(2*N-1)     :: IND
12  INTEGER,DIMENSION(N+1)       :: IA_AUX,IAR
13  REAL(8),ALLOCATABLE          :: SA(:),COEF(:,:),DIAG(:,:)
14  INTEGER,ALLOCATABLE          :: IJA(:),JCOEF(:,:),IOFF(:)
15  REAL(8),DIMENSION(N)         :: X_OLD,X,P_OLD,P,R_OLD,R,RS,
        Q,Q_OLD,UHAT,V,PHAT,U,QHAT
16  REAL(8)                      :: SUM_RF,ARF,RF,RFSN,RFS,NORM,
        ALPHA,BETA,TIME_BEGIN,TIME_END,RHO,RHO_OLD
17  INTEGER,DIMENSION(N)         :: UPTR,IW
18  !---------------------------------------------
19  CALL TOTAL_LENGTH         (N,IA_AUX,TL)
20  ALLOCATE                  (SA(TL),IJA(TL))
21  CALL CSRMSR               (TL,N,NNZERO,AA_AUX,JA_AUX,IA_AUX,SA,
        IJA)
22  CALL INFDIA               (N,NNZERO,JA_AUX,IA_AUX,IND,IDIAG)
23  ALLOCATE                  (COEF(N,IDIAG),JCOEF(N,IDIAG))
24  ALLOCATE                  (IOFF(IDIAG),DIAG(N,IDIAG))
25  CALL CSRELL               (N,NNZERO,AA_AUX,JA_AUX,IA_AUX,N,COEF,
        JCOEF,N,IDIAG,IERR)
26  CALL CSRDIA               (N,NNZERO,IDIAG,10,AA_AUX,JA_AUX,
        IA_AUX,N,DIAG,IOFF,AAR,JAR,IAR,IND)
27  !---------------------------------------------
28  CALL ILU0    (N,NNZERO,AA_AUX,JA_AUX,IA_AUX,LUVAL,UPTR,IW,ICODE)
29  !---------------------------------------------
30  RS=R_OLD
31  NORM=1.D0
32  SUM_RF=0.D0
33  ITER=0
```

```
34   !CALL CPU_TIME (TIME_BEGIN)
35   DO WHILE (NORM.GT.TOL)
36       ITER=ITER+1
37       !PRINT*,ITER
38       RHO=DOT_PRODUCT(RS,R_OLD)
39   !-------------------------------------------------
40       IF (RHO.EQ.0.D0) PRINT*, 'conjugate gradient squared method
         fails'
41   !-------------------------------------------------
42       IF (ITER.EQ.1) THEN
43           U=R_OLD
44           P=U
45       ELSE
46           BETA=RHO/RHO_OLD
47           U=R_OLD+BETA*Q_OLD
48           P=U+BETA*(Q_OLD+BETA*P_OLD)
49       ENDIF
50   !-------------------------------------------------
51       CALL LUSOL   (N,NNZERO,P,PHAT,LUVAL,JA_AUX,IA_AUX,UPTR)
52   !-------------------------------------------------
53   ! Stores the matrix in compressed spars row and multiplies by
54   ! vector
55   !-------------------------------------------------
56       CALL CSR_MAT_V_PRODUCT (N,NNZERO,AA_AUX,JA_AUX,IA_AUX,PHAT,V)
57   !-------------------------------------------------
58   ! Stores the matrix in modified spars row and multiplies by
59   ! vector
60   !-------------------------------------------------
61   ! CALL MSR_MAT_V_PRODUCT (TL,N,SA,IJA,PHAT,V)
62   !-------------------------------------------------
63   ! Stores the matrix in Ellpack/Itpack format and multiplies by
64   ! vector
65   !-------------------------------------------------
66   ! CALL ELLPACK_ITPACK_MAT_V_PRODUCT (N,PHAT,V,N,IDIAG,COEF,JCOEF)
67   !-------------------------------------------------
68   ! ! Stores the matrix in Diagonal format and multiplies by
69   ! vector
70   !-------------------------------------------------
71   ! CALL DIAGONAL_MAT_V_PRODUCT (N,PHAT,V,DIAG,N,IDIAG,IOFF)
72   !-------------------------------------------------
73       ALPHA=RHO/(DOT_PRODUCT(RS,V))
74       Q=U-ALPHA*V
75       CALL LUSOL   (N,NNZERO,U+Q,UHAT,LUVAL,JA_AUX,IA_AUX,UPTR)
76       X=X_OLD+ALPHA*UHAT
77   !-------------------------------------------------
78   ! Stores the matrix in compressed spars row and multiplies by
79   ! vector
80   !-------------------------------------------------
81       CALL CSR_MAT_V_PRODUCT    (N,NNZERO,AA_AUX,JA_AUX,IA_AUX,UHAT,
         QHAT)
82   !-------------------------------------------------
83   ! Stores the matrix in modified spars row and multiplies by
84   ! vector
85   !-------------------------------------------------
86   ! CALL MSR_MAT_V_PRODUCT (TL,N,SA,IJA,UHAT,QHAT)
87   !-------------------------------------------------
88   ! Stores the matrix in Ellpack/Itpack format and multiplies by
89   ! vector
90   !-------------------------------------------------
91   ! CALL ELLPACK_ITPACK_MAT_V_PRODUCT (N,UHAT,QHAT,N,IDIAG,COEF,
         JCOEF)
92   !-------------------------------------------------
93   ! ! Stores the matrix in Diagonal format and multiplies by
```

```
 94  ! vector
 95  !--------------------------------------------------------
 96  ! CALL DIAGONAL_MAT_V_PRODUCT (N,UHAT,QHAT,DIAG,N,IDIAG,IOFF)
 97  !--------------------------------------------------------
 98     R=R_OLD-ALPHA*QHAT
 99     NORM=DSQRT(DOT_PRODUCT(R,R))/N
100  !--------------------------------------------------------
101  ! Computing the reduction factor
102  !--------------------------------------------------------
103     RFS=DOT_PRODUCT(R_OLD,R_OLD)
104     RFSN=DSQRT(RFS)/N
105     RF=NORM/RFSN
106     SUM_RF=SUM_RF+RF
107  !--------------------------------------------------------
108     P_OLD=P
109     X_OLD=X
110     R_OLD=R
111     Q_OLD=Q
112     RHO_OLD=RHO
113  !PRINT*,NORM
114  !CALL CPU_TIME (TIME_END)
115  !CALL CPU_TIME_WRITE (TIME_BEGIN,TIME_END,ITER)
116  !CALL SUCCESSIVE_SOL (NORM,ITER)
117  END DO
118  !ARF=SUM_RF/ITER
119  !PRINT*,ARF
120  !--------------------------------------------------------
121  END SUBROUTINE ILU0_PCGS
```

C.3   ILU(0)-Preconditioned Bi-Conjugate Gradient Stabilized Method

```
 1  !******************************************************************
 2  !!ILU(0)-Preconditioned Bi-Conjugate Gradient Stabilized Method
 3  !******************************************************************
 4  SUBROUTINE ILU0_PBCGSTAB (N,NNZERO,AA_AUX,JA_AUX,IA_AUX,X_OLD,
         R_OLD,X)
 5  !--------------------------------------------------
 6  USE PARAM, ONLY                    : TOL
 7  IMPLICIT NONE
 8  INTEGER                            :: ITER,N,NNZERO,TL,ICODE,
        IDIAG,IERR
 9  REAL(8),DIMENSION(NNZERO)          :: AA_AUX,AAR,LUVAL
10  INTEGER,DIMENSION(NNZERO)          :: JA_AUX,JAR
11  INTEGER,DIMENSION(2*N-1)           :: IND
12  INTEGER,DIMENSION(N+1)             :: IA_AUX,IAR
13  REAL(8),ALLOCATABLE                :: SA(:),COEF(:,:),DIAG(:,:)
14  INTEGER,ALLOCATABLE                :: IJA(:),JCOEF(:,:),IOFF(:)
15  REAL(8),DIMENSION(N)               :: X_OLD,X,P_OLD,P,R_OLD,R,RS,
        SR,SRHAT,T,V,V_OLD,PHAT
16  REAL(8)                            :: SUM_RF,ARF,RF,RFSN,RFS,
        NORM,TIME_BEGIN,TIME_END,OM,OM_OLD,ALPHA,ALPHA_OLD,BETA,RHO,
        RHO_OLD
17  INTEGER,DIMENSION(N)               :: UPTR,IW
18  !--------------------------------------------------
19  CALL TOTAL_LENGTH              (N,IA_AUX,TL)
20  ALLOCATE                       (SA(TL),IJA(TL))
```

```
21  CALL CSRMSR                (TL,N,NNZERO,AA_AUX,JA_AUX,IA_AUX,SA,
        IJA)
22  CALL INFDIA                (N,NNZERO,JA_AUX,IA_AUX,IND,IDIAG)
23  ALLOCATE                   (COEF(N,IDIAG),JCOEF(N,IDIAG))
24  ALLOCATE                   (IOFF(IDIAG),DIAG(N,IDIAG))
25  CALL CSRELL                (N,NNZERO,AA_AUX,JA_AUX,IA_AUX,N,COEF,
        JCOEF,N,IDIAG,IERR)
26  CALL CSRDIA                (N,NNZERO,IDIAG,10,AA_AUX,JA_AUX,
        IA_AUX,N,DIAG,IOFF,AAR,JAR,IAR,IND)
27  !-----------------------------------------------
28  CALL ILU0    (N,NNZERO,AA_AUX,JA_AUX,IA_AUX,LUVAL,UPTR,IW,ICODE)
29  !-----------------------------------------------
30  RS=R_OLD
31  NORM=1.D0
32  !SUM_RF=0.D0
33  ITER=0
34  !CALL CPU_TIME (TIME_BEGIN)
35  DO WHILE (NORM.GT.TOL)
36     ITER=ITER+1
37     !PRINT*,ITER
38     RHO=DOT_PRODUCT(RS,R_OLD)
39  !-----------------------------------------------
40     IF (RHO.EQ.0.D0) PRINT*, 'ILU0_PBiconjugate gradient
        stabilized method fails'
41  !-----------------------------------------------
42     IF (ITER.EQ.1) THEN
43        P=R_OLD
44     ELSE
45        BETA=(RHO/RHO_OLD)*(ALPHA_OLD/OM_OLD)
46        P=R_OLD+BETA*(P_OLD-OM_OLD*V_OLD)
47     END IF
48  !-----------------------------------------------
49     CALL LUSOL    (N,NNZERO,P,PHAT,LUVAL,JA_AUX,IA_AUX,UPTR)
50  !-----------------------------------------------
51  ! Stores the matrix in compressed spars row and multiplies by
52  ! vector
53  !-----------------------------------------------
54     CALL CSR_MAT_V_PRODUCT    (N,NNZERO,AA_AUX,JA_AUX,IA_AUX,PHAT,
        V)
55  !-----------------------------------------------
56  ! Stores the matrix in modified spars row and multiplies by
57  ! vector
58  !-----------------------------------------------
59  ! CALL MSR_MAT_V_PRODUCT (TL,N,SA,IJA,PHAT,V)
60  !-----------------------------------------------
61  ! Stores the matrix in Ellpack/Itpack format and multiplies by
62  ! vector
63  !-----------------------------------------------
64  ! CALL ELLPACK_ITPACK_MAT_V_PRODUCT (N,PHAT,V,N,IDIAG,COEF,JCOEF)
65  !-----------------------------------------------
66  ! Stores the matrix in Diagonal format and multiplies by
67  ! vector
68  !-----------------------------------------------
69  ! CALL DIAGONAL_MAT_V_PRODUCT (N,PHAT,V,DIAG,N,IDIAG,IOFF)
70  !-----------------------------------------------
71     ALPHA=RHO/DOT_PRODUCT(RS,V)
72     SR=R_OLD-ALPHA*V
73  !-----------------------------------------------
74     CALL LUSOL    (N,NNZERO,SR,SRHAT,LUVAL,JA_AUX,IA_AUX,UPTR)
75  !-----------------------------------------------
76  ! Stores the matrix in compressed spars row and multiplies by
77  ! vector
78  !-----------------------------------------------
```

```
79      CALL CSR_MAT_V_PRODUCT        (N,NNZERO,AA_AUX,JA_AUX,IA_AUX,
        SRHAT,T)
80  !-----------------------------------------------------
81  ! Stores the matrix in modified spars row and multiplies by
82  ! vector
83  !-----------------------------------------------------
84  ! CALL MSR_MAT_V_PRODUCT (TL,N,SA,IJA,SRHAT,T)
85  !-----------------------------------------------------
86  ! Stores the matrix in Ellpack/Itpack format and multiplies by
87  ! vector
88  !-----------------------------------------------------
89  ! CALL ELLPACK_ITPACK_MAT_V_PRODUCT (N,SRHAT,T,N,IDIAG,COEF,JCOEF)
90  !-----------------------------------------------------
91  ! Stores the matrix in Diagonal format and multiplies by
92  ! vector
93  !-----------------------------------------------------
94  ! CALL DIAGONAL_MAT_V_PRODUCT (N,SRHAT,T,DIAG,N,IDIAG,IOFF)
95  !-----------------------------------------------------
96      OM=DOT_PRODUCT(T,SR)/DOT_PRODUCT(T,T)
97      X= X_OLD+ALPHA*PHAT+OM*SRHAT
98      R=SR-OM*T
99      NORM=DSQRT(DOT_PRODUCT(R,R))/N
100     !PRINT*,NORM
101 !-----------------------------------------------------
102 ! Computing the reduction factor
103 !-----------------------------------------------------
104 !RFS=DOT_PRODUCT(R_OLD,R_OLD)
105 !RFSN=DSQRT(RFS)/N
106 !RF=NORM/RFSN
107 !SUM_RF=SUM_RF+RF
108 !-----------------------------------------------------
109     X_OLD=X
110     R_OLD=R
111     RHO_OLD=RHO
112     P_OLD=P
113     OM_OLD=OM
114     V_OLD=V
115     ALPHA_OLD=ALPHA
116 !CALL CPU_TIME (TIME_END)
117 !CALL CPU_TIME_WRITE (TIME_BEGIN,TIME_END,ITER)
118 !CALL SUCCESSIVE_SOL (NORM,ITER)
119 END DO
120 !ARF=SUM_RF/ITER
121 !PRINT*,ARF
122 !-----------------------------------------------------
123 END SUBROUTINE ILU0_PBCGSTAB
```

Appendix D

Inner Iterations of GMRES* Method

D.1 Conjugate Gradient Method Inner Iterations

Subroutines of this appendix have a similar argument list and nested calls. Therefore, we will provide explanations in this section and refrain from duplicating the content in other sections. On entry of the argument list of the subroutine, MD and NNZERO denote the matrix dimension and number of non-zero elements. INNER_ITER is the number of inner iterations of the CG method. The matrix is presented via three vectors AA_AUX, JA_AUX and IA_AUX in CSR format. X_OLD is the initial guess or the approximate solution from the exterior iterations of the GMRES method. B is the initial residual or the residual of the solution from the exterior iterations of the GMRES method. On return, X is the approximate solution vector after INNER_ITER iterations. For more details refer to Algorithm 3.7.

```
1  !***************************************************
2  !   Conjugate Gradien Method Inner Iteration
3  !***************************************************
4  SUBROUTINE CG_INNER (MD,NNZERO,INNER_ITER,AA_AUX,JA_AUX,IA_AUX,X_OLD,
       B,X)
5  !-------------------------------------------------
6  USE PARAM, ONLY             : TOL
7  IMPLICIT NONE
8  INTEGER                     :: K,TL,MD,NNZERO,IDIAG,IERR,I,
       INNER_ITER
9  REAL(8),DIMENSION(MD)       :: X_OLD,X,P_OLD,P,R_OLD,R,AP,B
10 INTEGER,DIMENSION(MD+1)     :: IA_AUX,IAR
11 INTEGER,DIMENSION(2*MD-1)   :: IND
12 REAL(8)                     :: S,ALPHA,BETA,M,MM,TIME_BEGIN,
       TIME_END
13 REAL(8),DIMENSION(NNZERO)   :: AA_AUX,AAR
14 INTEGER,DIMENSION(NNZERO)   :: JA_AUX,JAR
15 REAL(8),ALLOCATABLE         :: SA(:),DIAG(:,:),COEF(:,:)
16 INTEGER,ALLOCATABLE         :: IJA(:),IOFF(:),JCOEF(:,:)
```

Krylov Subspace Methods with Application in Incompressible Fluid Flow Solvers, First Edition.
Iman Farahbakhsh.
© 2020 John Wiley & Sons Ltd. Published 2020 by John Wiley & Sons Ltd.
Companion Website: www.wiley.com/go/Farahbakhs/KrylovSubspaceMethods

```
17  !--------------------------------------------------
18  CALL TOTAL_LENGTH            (MD,IA_AUX,TL)
19  ALLOCATE                     (SA(TL),IJA(TL))
20  CALL CSRMSR                  (TL,MD,NNZERO,AA_AUX,JA_AUX,IA_AUX,SA,IJA)
21  CALL INFDIA                  (MD,NNZERO,JA_AUX,IA_AUX,IND,IDIAG)
22  ALLOCATE                     (COEF(MD,IDIAG),JCOEF(MD,IDIAG))
23  ALLOCATE                     (IOFF(IDIAG),DIAG(MD,IDIAG))
24  CALL CSRELL                  (MD,NNZERO,AA_AUX,JA_AUX,IA_AUX,MD,COEF,
         JCOEF,MD,IDIAG,IERR)
25  CALL CSRDIA                  (MD,NNZERO,IDIAG,10,AA_AUX,JA_AUX,IA_AUX,
         MD,DIAG,IOFF,AAR,JAR,IAR,IND)
26  !--------------------------------------------------
27  CALL INITIAL_RESIDUAL (MD,NNZERO,AA_AUX,JA_AUX,IA_AUX,B,X_OLD,R_OLD)
28  !--------------------------------------------------
29  P_OLD=R_OLD
30  DO I=1,INNER_ITER
31     S=0.D0
32     M=0.D0
33     MM=0.D0
34  !--------------------------------------------------
35  ! Stores the matrix in compressed spars row and multiplies by
36  ! vector
37  !--------------------------------------------------
38  !CALL CSR_MAT_V_PRODUCT (MD,NNZERO,AA_AUX,JA_AUX,IA_AUX,P_OLD,AP)
39  !--------------------------------------------------
40  ! Stores the matrix in modified spars row and multiplies by
41  ! vector
42  !--------------------------------------------------
43     CALL MSR_MAT_V_PRODUCT (TL,MD,SA,IJA,P_OLD,AP)
44  !--------------------------------------------------
45  ! Stores the matrix in Ellpack/Itpack format and multiplies by
46  ! vector
47  !--------------------------------------------------
48  !CALL ELLPACK_ITPACK_MAT_V_PRODUCT (MD,P_OLD,AP,MD,IDIAG,COEF,JCOEF)
49  !--------------------------------------------------
50  ! Stores the matrix in Diagonal format and multiplies by
51  ! vector
52  !--------------------------------------------------
53  !CALL  DIAGONAL_MAT_V_PRODUCT (MD,P_OLD,AP,DIAG,MD,IDIAG,IOFF)
54  !--------------------------------------------------
55     DO K=1,MD
56        M=M+R_OLD(K)**2
57        MM=MM+AP(K)*P_OLD(K)
58     END DO
59     ALPHA=M/MM
60     X=X_OLD+ALPHA*P_OLD
61     R=R_OLD-ALPHA*AP
62     DO K=1,MD
63        S=S+R(K)**2
64     END DO
65     BETA=S/M
66     P=R+BETA*P_OLD
67     P_OLD=P
68     X_OLD=X
69     R_OLD=R
70  END DO
71  !--------------------------------------------------
72  END SUBROUTINE CG_INNER
```

D.2   Conjugate Gradient Squared Method Inner Iterations

```
1  !****************************************************
2  !Conjugate Gradien Squared Method Inner Iteration
3  !****************************************************
4  SUBROUTINE CGS_INNER (N,NNZERO,INNER_ITER,AA_AUX,JA_AUX,IA_AUX,X_OLD,
       B,X)
5  !-------------------------------------------------
6  USE PARAM, ONLY              : TOL
7  IMPLICIT NONE
8  INTEGER                      :: I,K,N,NNZERO,TL,IDIAG,IERR,
       INNER_ITER
9  REAL(8),DIMENSION(NNZERO)    :: AA_AUX,AAR
10 INTEGER,DIMENSION(NNZERO)    :: JA_AUX,JAR
11 INTEGER,DIMENSION(2*N-1)     :: IND
12 INTEGER,DIMENSION(N+1)       :: IA_AUX,IAR
13 REAL(8),ALLOCATABLE          :: SA(:),COEF(:,:),DIAG(:,:)
14 INTEGER,ALLOCATABLE          :: IJA(:),JCOEF(:,:),IOFF(:)
15 REAL(8),DIMENSION(N)         :: X_OLD,X,P_OLD,P,R_OLD,U,U_OLD,R,
       AP,AUQ,RS,Q,B
16 REAL(8)                      :: ALPHA,BETA,M,MM,MN,TIME_BEGIN,
       TIME_END
17 !-------------------------------------------------
18 CALL TOTAL_LENGTH           (N,IA_AUX,TL)
19 ALLOCATE                    (SA(TL),IJA(TL))
20 CALL CSRMSR                 (TL,N,NNZERO,AA_AUX,JA_AUX,IA_AUX,SA,IJA)
21 CALL INFDIA                 (N,NNZERO,JA_AUX,IA_AUX,IND,IDIAG)
22 ALLOCATE                    (COEF(N,IDIAG),JCOEF(N,IDIAG))
23 ALLOCATE                    (IOFF(IDIAG),DIAG(N,IDIAG))
24 CALL CSRELL                 (N,NNZERO,AA_AUX,JA_AUX,IA_AUX,N,COEF,JCOEF,
       N,IDIAG,IERR)
25 CALL CSRDIA                 (N,NNZERO,IDIAG,10,AA_AUX,JA_AUX,IA_AUX,N,
       DIAG,IOFF,AAR,JAR,IAR,IND)
26 !-------------------------------------------------
27 CALL INITIAL_RESIDUAL (N,NNZERO,AA_AUX,JA_AUX,IA_AUX,B,X_OLD,R_OLD)
28 !-------------------------------------------------
29 RS=R_OLD
30 P_OLD=R_OLD
31 U_OLD=R_OLD
32 DO I=1,INNER_ITER
33    M=0.D0
34    MM=0.D0
35    MN=0.D0
36 !-------------------------------------------------
37 ! Stores the matrix in compressed spars row and multiplies by
38 ! vector
39 !-------------------------------------------------
40 !CALL CSR_MAT_V_PRODUCT (N,NNZERO,AA_AUX,JA_AUX,IA_AUX,P_OLD,AP)
41 !-------------------------------------------------
42 ! Stores the matrix in modified spars row and multiplies by
43 ! vector
44 !-------------------------------------------------
45 !CALL MSR_MAT_V_PRODUCT (TL,N,SA,IJA,P_OLD,AP)
46 !-------------------------------------------------
47 ! Stores the matrix in Ellpack/Itpack format and multiplies by
48 ! vector
49 !-------------------------------------------------
50 !CALL ELLPACK_ITPACK_MAT_V_PRODUCT (N,P_OLD,AP,N,IDIAG,COEF,JCOEF)
51 !-------------------------------------------------
52 ! Stores the matrix in Diagonal format and multiplies by
53 ! vector
54 !-------------------------------------------------
55    CALL DIAGONAL_MAT_V_PRODUCT (N,P_OLD,AP,DIAG,N,IDIAG,IOFF)
```

```
56  !----------------------------------------------------------
57     DO K=1,N
58        M=M+R_OLD(K)*RS(K)
59        MM=MM+AP(K)*RS(K)
60     END DO
61     ALPHA=M/MM
62     Q=U_OLD-ALPHA*AP
63     X=X_OLD+ALPHA*(U_OLD+Q)
64  !----------------------------------------------------------
65  ! Stores the matrix in compressed spars row and multiplies by
66  ! vector
67  !----------------------------------------------------------
68  !CALL CSR_MAT_V_PRODUCT    (N,NNZERO,AA_AUX,JA_AUX,IA_AUX,U_OLD+Q,AUQ)
69  !----------------------------------------------------------
70  ! Stores the matrix in modified spars row and multiplies by
71  ! vector
72  !----------------------------------------------------------
73  !CALL MSR_MAT_V_PRODUCT (TL,N,SA,IJA,U_OLD+Q,AUQ)
74  !----------------------------------------------------------
75  ! Stores the matrix in Ellpack/Itpack format and multiplies by
76  ! vector
77  !----------------------------------------------------------
78  !CALL ELLPACK_ITPACK_MAT_V_PRODUCT (N,U_OLD+Q,AUQ,N,IDIAG,COEF,JCOEF)
79  !----------------------------------------------------------
80  ! Stores the matrix in Diagonal format and multiplies by
81  ! vector
82  !----------------------------------------------------------
83     CALL DIAGONAL_MAT_V_PRODUCT (N,U_OLD+Q,AUQ,DIAG,N,IDIAG,IOFF)
84  !----------------------------------------------------------
85     R=R_OLD-ALPHA*AUQ
86     DO K=1,N
87        MN=MN+R(K)*RS(K)
88     END DO
89     BETA=MN/M
90     U=R+BETA*Q
91     P=U+BETA*(Q+BETA*P_OLD)
92     P_OLD=P
93     X_OLD=X
94     R_OLD=R
95     U_OLD=U
96  END DO
97  !----------------------------------------------------------
98  END SUBROUTINE CGS_INNER
```

D.3   Bi-Conjugate Gradient Stabilized Method Inner Iterations

```
1   !**************************************************************
2   !Bi-Conjugate Gradient Stabilized Method Inner Iteration
3   !**************************************************************
4   SUBROUTINE BCGSTAB_INNER (N,NNZERO,INNER_ITER,AA_AUX,JA_AUX,IA_AUX,
        X_OLD,B,X)
5   !----------------------------------------------------------
6   USE PARAM, ONLY              : TOL
7   IMPLICIT NONE
8   INTEGER                      :: I,K,N,NNZERO,TL,IDIAG,IERR,
        INNER_ITER
9   REAL(8),DIMENSION(NNZERO)    :: AA_AUX,AAR
10  INTEGER,DIMENSION(NNZERO)    :: JA_AUX,JAR
11  INTEGER,DIMENSION(2*N-1)     :: IND
12  INTEGER,DIMENSION(N+1)       :: IA_AUX,IAR
```

```
13  REAL(8),ALLOCATABLE                :: SA(:),COEF(:,:),DIAG(:,:)
14  INTEGER,ALLOCATABLE                :: IJA(:),JCOEF(:,:),IOFF(:)
15  REAL(8),DIMENSION(N)               :: X_OLD,X,P_OLD,P,R_OLD,R,AP,AQQ,
        RS,QQ,B
16  REAL(8)                            :: ALPHA,BETA,M,MS,MMS,MM,MN,
        TIME_BEGIN,TIME_END,OM
17  !-----------------------------------------------------
18  CALL TOTAL_LENGTH          (N,IA_AUX,TL)
19  ALLOCATE                   (SA(TL),IJA(TL))
20  CALL CSRMSR                (TL,N,NNZERO,AA_AUX,JA_AUX,IA_AUX,SA,IJA)
21  CALL INFDIA                (N,NNZERO,JA_AUX,IA_AUX,IND,IDIAG)
22  ALLOCATE                   (COEF(N,IDIAG),JCOEF(N,IDIAG))
23  ALLOCATE                   (IOFF(IDIAG),DIAG(N,IDIAG))
24  CALL CSRELL                (N,NNZERO,AA_AUX,JA_AUX,IA_AUX,N,COEF,JCOEF,
        N,IDIAG,IERR)
25  CALL CSRDIA                (N,NNZERO,IDIAG,10,AA_AUX,JA_AUX,IA_AUX,N,
        DIAG,IOFF,AAR,JAR,IAR,IND)
26  !-----------------------------------------------------
27  CALL INITIAL_RESIDUAL (N,NNZERO,AA_AUX,JA_AUX,IA_AUX,B,X_OLD,R_OLD)
28  !-----------------------------------------------------
29  RS=R_OLD
30  P_OLD=R_OLD
31  DO I=1,INNER_ITER
32     M=0.D0
33     MS=0.D0
34     MMS=0.D0
35     MM=0.D0
36     MN=0.D0
37  !-----------------------------------------------------
38  ! Stores the matrix in compressed spars row and multiplies by
39  ! vector
40  !-----------------------------------------------------
41  !CALL CSR_MAT_V_PRODUCT     (N,NNZERO,AA_AUX,JA_AUX,IA_AUX,P_OLD,AP)
42  !-----------------------------------------------------
43  ! Stores the matrix in modified spars row and multiplies by
44  ! vector
45  !-----------------------------------------------------
46    CALL MSR_MAT_V_PRODUCT (TL,N,SA,IJA,P_OLD,AP)
47  !-----------------------------------------------------
48  ! Stores the matrix in Ellpack/Itpack format and multiplies by
49  ! vector
50  !-----------------------------------------------------
51  !CALL ELLPACK_ITPACK_MAT_V_PRODUCT (N,P_OLD,AP,N,IDIAG,COEF,JCOEF)
52  !-----------------------------------------------------
53  ! Stores the matrix in Diagonal format and multiplies by
54  ! vector
55  !-----------------------------------------------------
56  !CALL DIAGONAL_MAT_V_PRODUCT (N,P_OLD,AP,DIAG,N,IDIAG,IOFF)
57  !-----------------------------------------------------
58     DO K=1,N
59        M=M+R_OLD(K)*RS(K)
60        MM=MM+AP(K)*RS(K)
61     END DO
62     ALPHA=M/MM
63     QQ=R_OLD-ALPHA*AP
64  !-----------------------------------------------------
65  ! Stores the matrix in compressed spars row and multiplies by
66  ! vector
67  !-----------------------------------------------------
68  !CALL CSR_MAT_V_PRODUCT     (N,NNZERO,AA_AUX,JA_AUX,IA_AUX,QQ,AQQ)
69  !-----------------------------------------------------
70  ! Stores the matrix in modified spars row and multiplies by
71  ! vector
72  !-----------------------------------------------------
73    CALL MSR_MAT_V_PRODUCT (TL,N,SA,IJA,QQ,AQQ)
74  !-----------------------------------------------------
75  ! Stores the matrix in Ellpack/Itpack format and multiplies by
```

```
 76  ! vector
 77  !----------------------------------------------------
 78  !CALL ELLPACK_ITPACK_MAT_V_PRODUCT (N,QQ,AQQ,N,IDIAG,COEF,JCOEF)
 79  !----------------------------------------------------
 80  ! Stores the matrix in Diagonal format and multiplies by
 81  ! vector
 82  !----------------------------------------------------
 83  !CALL DIAGONAL_MAT_V_PRODUCT (N,QQ,AQQ,DIAG,N,IDIAG,IOFF)
 84  !----------------------------------------------------
 85     DO K=1,N
 86        MS=MS+AQQ(K)*QQ(K)
 87        MMS=MMS+AQQ(K)*AQQ(K)
 88     END DO
 89     OM=MS/MMS
 90     X=X_OLD+ALPHA*P_OLD+OM*QQ
 91     R=QQ-OM*AQQ
 92     DO K=1,N
 93        MN=MN+R(K)*RS(K)
 94     END DO
 95     BETA=(MN/M)*(ALPHA/OM)
 96     P=R+BETA*(P_OLD-OM*AP)
 97     P_OLD=P
 98     X_OLD=X
 99     R_OLD=R
100  END DO
101  !----------------------------------------------------
102  END SUBROUTINE BCGSTAB_INNER
```

D.4   Conjugate Residual Method Inner Iterations

```
 1  !****************************************************
 2  !    Conjugate Residual Method Inner Iteration
 3  !****************************************************
 4  SUBROUTINE CR_INNER (MD,NNZERO,INNER_ITER,AA_AUX,JA_AUX,IA_AUX,X_OLD,
       B,X)
 5  !----------------------------------------------------
 6  USE PARAM, ONLY                  : TOL
 7  IMPLICIT NONE
 8  INTEGER                          :: TL,MD,NNZERO,IDIAG,IERR,
       INNER_ITER,I
 9  REAL(8),DIMENSION(MD)            :: X_OLD,X,P_OLD,P,R_OLD,R,AP,
       AP_OLD,AR_OLD,AR,B
10  INTEGER,DIMENSION(MD+1)          :: IA_AUX,IAR
11  INTEGER,DIMENSION(2*MD-1)        :: IND
12  REAL(8)                          :: ALPHA,BETA
13  REAL(8),DIMENSION(NNZERO)        :: AA_AUX,AAR
14  INTEGER,DIMENSION(NNZERO)        :: JA_AUX,JAR
15  REAL(8),ALLOCATABLE              :: SA(:),DIAG(:,:),COEF(:,:)
16  INTEGER,ALLOCATABLE              :: IJA(:),IOFF(:),JCOEF(:,:)
17  !----------------------------------------------------
18  CALL TOTAL_LENGTH       (MD,IA_AUX,TL)
19  ALLOCATE                (SA(TL),IJA(TL))
20  CALL CSRMSR             (TL,MD,NNZERO,AA_AUX,JA_AUX,IA_AUX,SA,IJA)
21  CALL INFDIA             (MD,NNZERO,JA_AUX,IA_AUX,IND,IDIAG)
22  ALLOCATE                (COEF(MD,IDIAG),JCOEF(MD,IDIAG))
23  ALLOCATE                (IOFF(IDIAG),DIAG(MD,IDIAG))
24  CALL CSRELL             (MD,NNZERO,AA_AUX,JA_AUX,IA_AUX,MD,COEF,
       JCOEF,MD,IDIAG,IERR)
25  CALL CSRDIA             (MD,NNZERO,IDIAG,10,AA_AUX,JA_AUX,IA_AUX,
       MD,DIAG,IOFF,AAR,JAR,IAR,IND)
26  !----------------------------------------------------
27  CALL INITIAL_RESIDUAL (MD,NNZERO,AA_AUX,JA_AUX,IA_AUX,B,X_OLD,R_OLD)
```

```
28  !--------------------------------------------------
29  P_OLD=R_OLD
30  CALL MSR_MAT_V_PRODUCT (TL,MD,SA,IJA,P_OLD,AP_OLD)
31  DO I=1,INNER_ITER
32  !--------------------------------------------------
33  ! Stores the matrix in compressed spars row and multiplies by
34  ! vector
35  !--------------------------------------------------
36  !CALL CSR_MAT_V_PRODUCT      (MD,NNZERO,AA_AUX,JA_AUX,IA_AUX,R_OLD,
         AR_OLD)
37  !--------------------------------------------------
38  ! Stores the matrix in modified spars row and multiplies by
39  ! vector
40  !--------------------------------------------------
41     CALL MSR_MAT_V_PRODUCT (TL,MD,SA,IJA,R_OLD,AR_OLD)
42  !--------------------------------------------------
43  ! Stores the matrix in Ellpack/Itpack format and multiplies by
44  ! vector
45  !--------------------------------------------------
46  !CALL ELLPACK_ITPACK_MAT_V_PRODUCT (MD,R_OLD,AR_OLD,MD,IDIAG,COEF,
         JCOEF)
47  !--------------------------------------------------
48  ! Stores the matrix in Diagonal format and multiplies by
49  ! vector
50  !--------------------------------------------------
51  !CALL DIAGONAL_MAT_V_PRODUCT (MD,R_OLD,AR_OLD,DIAG,MD,IDIAG,IOFF)
52  !--------------------------------------------------
53     ALPHA=DOT_PRODUCT(R_OLD,AR_OLD)/DOT_PRODUCT(AP_OLD,AP_OLD)
54     X=X_OLD+ALPHA*P_OLD
55     R=R_OLD-ALPHA*AP_OLD
56  !--------------------------------------------------
57  ! Stores the matrix in compressed spars row and multiplies by
58  ! vector
59  !--------------------------------------------------
60  !CALL CSR_MAT_V_PRODUCT      (MD,NNZERO,AA_AUX,JA_AUX,IA_AUX,R,AR)
61  !--------------------------------------------------
62  ! Stores the matrix in modified spars row and multiplies by
63  ! vector
64  !--------------------------------------------------
65     CALL MSR_MAT_V_PRODUCT (TL,MD,SA,IJA,R,AR)
66  !--------------------------------------------------
67  ! Stores the matrix in Ellpack/Itpack format and multiplies by
68  ! vector
69  !--------------------------------------------------
70  !CALL ELLPACK_ITPACK_MAT_V_PRODUCT (MD,R,AR,MD,IDIAG,COEF,JCOEF)
71  !--------------------------------------------------
72  ! Stores the matrix in Diagonal format and multiplies by
73  ! vector
74  !--------------------------------------------------
75  !CALL DIAGONAL_MAT_V_PRODUCT (MD,R,AR,DIAG,MD,IDIAG,IOFF)
76  !--------------------------------------------------
77     BETA=DOT_PRODUCT(R,AR)/DOT_PRODUCT(R_OLD,AR_OLD)
78     P=R+BETA*P_OLD
79     AP=AR+BETA*AP_OLD
80     P_OLD=P
81     X_OLD=X
82     R_OLD=R
83     AP_OLD=AP
84  END DO
85  !--------------------------------------------------
86  END SUBROUTINE CR_INNER
```

D.5 ILU(0) Preconditioned Conjugate Gradient Method Inner Iterations

```
1  !*****************************************************************
2  !ILU0 Preconditioned Conjugate Gradient Method Inner Iteration
3  !*****************************************************************
4  SUBROUTINE ILU0_PCG_INNER (N,NNZERO,INNER_ITER,AA_AUX,JA_AUX,IA_AUX,
       X_OLD,B,X)
5  !------------------------------------------------------
6  USE PARAM, ONLY              : TOL
7  IMPLICIT NONE
8  INTEGER                      :: K,I,TL,N,NNZERO,ICODE,IDIAG,IERR,
       INNER_ITER
9  REAL(8),DIMENSION(N)         :: X_OLD,X,P_OLD,P,R_OLD,R,AP,Z_OLD,
       Z,B
10 INTEGER,DIMENSION(N+1)       :: IA_AUX,IAR
11 INTEGER,DIMENSION(2*N-1)     :: IND
12 REAL(8)                      :: S,ALPHA,BETA,M,MM
13 REAL(8),DIMENSION(NNZERO)    :: AA_AUX,AAR,LUVAL
14 INTEGER,DIMENSION(NNZERO)    :: JA_AUX,JAR
15 REAL(8),ALLOCATABLE          :: SA(:),COEF(:,:),DIAG(:,:)
16 INTEGER,ALLOCATABLE          :: IJA(:),JCOEF(:,:),IOFF(:)
17 INTEGER,DIMENSION(N)         :: UPTR,IW
18 REAL(8)                      :: SN,TIME_BEGIN,TIME_END
19 !------------------------------------------------------
20 CALL TOTAL_LENGTH           (N,IA_AUX,TL)
21 ALLOCATE                    (SA(TL),IJA(TL))
22 CALL CSRMSR                 (TL,N,NNZERO,AA_AUX,JA_AUX,IA_AUX,SA,IJA)
23 CALL INFDIA                 (N,NNZERO,JA_AUX,IA_AUX,IND,IDIAG)
24 ALLOCATE                    (COEF(N,IDIAG),JCOEF(N,IDIAG))
25 ALLOCATE                    (IOFF(IDIAG),DIAG(N,IDIAG))
26 CALL CSRELL                 (N,NNZERO,AA_AUX,JA_AUX,IA_AUX,N,COEF,
       JCOEF,N,IDIAG,IERR)
27 CALL CSRDIA                 (N,NNZERO,IDIAG,10,AA_AUX,JA_AUX,IA_AUX,N,
       DIAG,IOFF,AAR,JAR,IAR,IND)
28 !------------------------------------------------------
29 CALL INITIAL_RESIDUAL (N,NNZERO,AA_AUX,JA_AUX,IA_AUX,B,X_OLD,R_OLD)
30 !------------------------------------------------------
31 CALL ILU0   (N,NNZERO,AA_AUX,JA_AUX,IA_AUX,LUVAL,UPTR,IW,ICODE)
32 CALL LUSOL  (N,NNZERO,R_OLD,Z_OLD,LUVAL,JA_AUX,IA_AUX,UPTR)
33 !------------------------------------------------------
34 P_OLD=Z_OLD
35 DO I=1,INNER_ITER
36     SN=0.D0
37     S=0.D0
38     M=0.D0
39     MM=0.D0
40 !------------------------------------------------------
41 ! Stores the matrix in compressed spars row and multiplies by
42 ! vector
43 !------------------------------------------------------
44 !CALL CSR_MAT_V_PRODUCT    (N,NNZERO,AA_AUX,JA_AUX,IA_AUX,P_OLD,AP)
45 !------------------------------------------------------
46 ! Stores the matrix in modified spars row and multiplies by
47 ! vector
48 !------------------------------------------------------
49 !CALL MSR_MAT_V_PRODUCT (TL,N,SA,IJA,P_OLD,AP)
50 !------------------------------------------------------
51 ! Stores the matrix in Ellpack/Itpack format and multiplies by
52 ! vector
53 !------------------------------------------------------
54 !CALL ELLPACK_ITPACK_MAT_V_PRODUCT (N,P_OLD,AP,N,IDIAG,COEF,JCOEF)
55 !------------------------------------------------------
56 ! Stores the matrix in Diagonal format and multiplies by
```

```
57  ! vector
58  !-----------------------------------------------------
59      CALL DIAGONAL_MAT_V_PRODUCT (N,P_OLD,AP,DIAG,N,IDIAG,IOFF)
60  !-----------------------------------------------------
61      DO K=1,N
62        M=M+R_OLD(K)*Z_OLD(K)
63        MM=MM+AP(K)*P_OLD(K)
64      END DO
65      ALPHA=M/MM
66      X=X_OLD+ALPHA*P_OLD
67      R=R_OLD-ALPHA*AP
68  !-----------------------------------------------------
69      CALL LUSOL    (N,NNZERO,R,Z,LUVAL,JA_AUX,IA_AUX,UPTR)
70  !-----------------------------------------------------
71      DO K=1,N
72        S=S+R(K)*Z(K)
73      END DO
74      BETA=S/M
75      P=Z+BETA*P_OLD
76      P_OLD=P
77      X_OLD=X
78      R_OLD=R
79      Z_OLD=Z
80  END DO
81  !-----------------------------------------------------
82  END SUBROUTINE ILU0_PCG_INNER
```

D.6   ILU(0) Preconditioned Conjugate Gradient Squared Method Inner Iterations

```
1   !***********************************************************************
2   !ILU0 Preconditioned Conjugate Gradient Squared Method Inner Iteration
3   !***********************************************************************
4   SUBROUTINE ILU0_PCGS_INNER (N,NNZERO,INNER_ITER,AA_AUX,JA_AUX,IA_AUX,
        X_OLD,B,X)
5   !-------------------------------------------
6   USE PARAM, ONLY              : TOL
7   IMPLICIT NONE
8   INTEGER                      :: I,N,NNZERO,TL,ICODE,IDIAG,IERR,
        INNER_ITER
9   REAL(8),DIMENSION(NNZERO)    :: AA_AUX,AAR,LUVAL
10  INTEGER,DIMENSION(NNZERO)    :: JA_AUX,JAR
11  INTEGER,DIMENSION(2*N-1)     :: IND
12  INTEGER,DIMENSION(N+1)       :: IA_AUX,IAR
13  REAL(8),ALLOCATABLE          :: SA(:),COEF(:,:),DIAG(:,:)
14  INTEGER,ALLOCATABLE          :: IJA(:),JCOEF(:,:),IOFF(:)
15  REAL(8),DIMENSION(N)         :: X_OLD,X,P_OLD,P,R_OLD,R,RS,Q,
        Q_OLD,UHAT,V,PHAT,U,QHAT,B
16  REAL(8)                      :: ALPHA,BETA,TIME_BEGIN,TIME_END,
        RHO,RHO_OLD
17  INTEGER,DIMENSION(N)         :: UPTR,IW
18  !-------------------------------------------
19  CALL TOTAL_LENGTH           (N,IA_AUX,TL)
20  ALLOCATE                    (SA(TL),IJA(TL))
21  CALL CSRMSR                 (TL,N,NNZERO,AA_AUX,JA_AUX,IA_AUX,SA,IJA)
22  CALL INFDIA                 (N,NNZERO,JA_AUX,IA_AUX,IND,IDIAG)
23  ALLOCATE                    (COEF(N,IDIAG),JCOEF(N,IDIAG))
24  ALLOCATE                    (IOFF(IDIAG),DIAG(N,IDIAG))
25  CALL CSRELL                 (N,NNZERO,AA_AUX,JA_AUX,IA_AUX,N,COEF,
        JCOEF,N,IDIAG,IERR)
```

```
26  CALL CSRDIA                    (N,NNZERO,IDIAG,10,AA_AUX,JA_AUX,IA_AUX,N,
        DIAG,IOFF,AAR,JAR,IAR,IND)
27  !-----------------------------------------------
28  CALL INITIAL_RESIDUAL (N,NNZERO,AA_AUX,JA_AUX,IA_AUX,B,X_OLD,R_OLD)
29  !-----------------------------------------------
30  CALL ILU0    (N,NNZERO,AA_AUX,JA_AUX,IA_AUX,LUVAL,UPTR,IW,ICODE)
31  !-----------------------------------------------
32  RS=R_OLD
33  DO I=1,INNER_ITER
34     RHO=DOT_PRODUCT(RS,R_OLD)
35  !-----------------------------------------------
36     IF (RHO.EQ.0.D0) PRINT*, 'conjugate gradient squared method fails'
37  !-----------------------------------------------
38     IF (I.EQ.1) THEN
39        U=R_OLD
40        P=U
41     ELSE
42        BETA=RHO/RHO_OLD
43        U=R_OLD+BETA*Q_OLD
44        P=U+BETA*(Q_OLD+BETA*P_OLD)
45     ENDIF
46  !-----------------------------------------------
47     CALL LUSOL    (N,NNZERO,P,PHAT,LUVAL,JA_AUX,IA_AUX,UPTR)
48  !-----------------------------------------------
49  !Stores the matrix in compressed spars row and multiplies by
50  !vector
51  !-----------------------------------------------
52  !CALL CSR_MAT_V_PRODUCT       (N,NNZERO,AA_AUX,JA_AUX,IA_AUX,PHAT,V)
53  !-----------------------------------------------
54  !Stores the matrix in modified spars row and multiplies by
55  !vector
56  !-----------------------------------------------
57     CALL MSR_MAT_V_PRODUCT (TL,N,SA,IJA,PHAT,V)
58  !-----------------------------------------------
59  !Stores the matrix in Ellpack/Itpack format and multiplies by
60  !vector
61  !-----------------------------------------------
62  !CALL ELLPACK_ITPACK_MAT_V_PRODUCT (N,PHAT,V,N,IDIAG,COEF,JCOEF)
63  !-----------------------------------------------
64  !Stores the matrix in Diagonal format and multiplies by
65  !vector
66  !-----------------------------------------------
67  !CALL DIAGONAL_MAT_V_PRODUCT (N,PHAT,V,DIAG,N,IDIAG,IOFF)
68  !-----------------------------------------------
69     ALPHA=RHO/(DOT_PRODUCT(RS,V))
70     Q=U-ALPHA*V
71     CALL LUSOL    (N,NNZERO,U+Q,UHAT,LUVAL,JA_AUX,IA_AUX,UPTR)
72     X=X_OLD+ALPHA*UHAT
73  !-----------------------------------------------
74  !Stores the matrix in compressed spars row and multiplies by
75  !vector
76  !-----------------------------------------------
77  !CALL CSR_MAT_V_PRODUCT       (N,NNZERO,AA_AUX,JA_AUX,IA_AUX,UHAT,QHAT)
78  !-----------------------------------------------
79  !Stores the matrix in modified spars row and multiplies by
80  !vector
81  !-----------------------------------------------
82     CALL MSR_MAT_V_PRODUCT (TL,N,SA,IJA,UHAT,QHAT)
83  !-----------------------------------------------
84  !Stores the matrix in Ellpack/Itpack format and multiplies by
85  !vector
86  !-----------------------------------------------
87  !CALL ELLPACK_ITPACK_MAT_V_PRODUCT (N,UHAT,QHAT,N,IDIAG,COEF,JCOEF)
88  !-----------------------------------------------
89  !Stores the matrix in Diagonal format and multiplies by
90  !vector
91  !-----------------------------------------------
```

```
92  !CALL DIAGONAL_MAT_V_PRODUCT (N,UHAT,QHAT,DIAG,N,IDIAG,IOFF)
93  !----------------------------------------------------
94     R=R_OLD-ALPHA*QHAT
95     P_OLD=P
96     X_OLD=X
97     R_OLD=R
98     Q_OLD=Q
99     RHO_OLD=RHO
100 END DO
101 !----------------------------------------------------
102 END SUBROUTINE ILU0_PCGS_INNER
```

D.7 ILU(0) Preconditioned Bi-Conjugate Gradient Stabilized Method Inner Iterations

```
1   !******************************************************
2   !ILU0 Preconditioned Bi-Conjugate Gradient
3   !Stabilized Method Inner Iteration
4   !******************************************************
5   SUBROUTINE ILU0_PBCGSTAB_INNER (N,NNZERO,INNER_ITER,AA_AUX,JA_AUX,
         IA_AUX,X_OLD,B,X)
6   !----------------------------------------------------
7   USE PARAM, ONLY                  : TOL
8   IMPLICIT NONE
9   INTEGER                          :: I,N,NNZERO,TL,ICODE,IDIAG,IERR,
         INNER_ITER
10  REAL(8),DIMENSION(NNZERO)        :: AA_AUX,AAR,LUVAL
11  INTEGER,DIMENSION(NNZERO)        :: JA_AUX,JAR
12  INTEGER,DIMENSION(2*N-1)         :: IND
13  INTEGER,DIMENSION(N+1)           :: IA_AUX,IAR
14  REAL(8),ALLOCATABLE              :: SA(:),COEF(:,:),DIAG(:,:)
15  INTEGER,ALLOCATABLE              :: IJA(:),JCOEF(:,:),IOFF(:)
16  REAL(8),DIMENSION(N)             :: X_OLD,X,P_OLD,P,R_OLD,R,RS,SR,
         SRHAT,T,V,V_OLD,PHAT,B
17  REAL(8)                          :: TIME_BEGIN,TIME_END,OM,OM_OLD,
         ALPHA,ALPHA_OLD,BETA,RHO,RHO_OLD
18  INTEGER,DIMENSION(N)             :: UPTR,IW
19  !----------------------------------------------------
20  CALL TOTAL_LENGTH        (N,IA_AUX,TL)
21  ALLOCATE                 (SA(TL),IJA(TL))
22  CALL CSRMSR              (TL,N,NNZERO,AA_AUX,JA_AUX,IA_AUX,SA,IJA)
23  CALL INFDIA              (N,NNZERO,JA_AUX,IA_AUX,IND,IDIAG)
24  ALLOCATE                 (COEF(N,IDIAG),JCOEF(N,IDIAG))
25  ALLOCATE                 (IOFF(IDIAG),DIAG(N,IDIAG))
26  CALL CSRELL              (N,NNZERO,AA_AUX,JA_AUX,IA_AUX,N,COEF,
         JCOEF,N,IDIAG,IERR)
27  CALL CSRDIA              (N,NNZERO,IDIAG,10,AA_AUX,JA_AUX,IA_AUX,N,
         DIAG,IOFF,AAR,JAR,IAR,IND)
28  !----------------------------------------------------
29  CALL INITIAL_RESIDUAL (N,NNZERO,AA_AUX,JA_AUX,IA_AUX,B,X_OLD,R_OLD)
30  !----------------------------------------------------
31  CALL ILU0     (N,NNZERO,AA_AUX,JA_AUX,IA_AUX,LUVAL,UPTR,IW,ICODE)
32  !----------------------------------------------------
33  RS=R_OLD
34  DO I=1,INNER_ITER
35     RHO=DOT_PRODUCT(RS,R_OLD)
36  !----------------------------------------------------
37     IF (RHO.EQ.0.D0) PRINT*, 'Biconjugate gradient stabilized method
         fails'
38  !----------------------------------------------------
39     IF (I.EQ.1) THEN
```

```
40          P=R_OLD
41       ELSE
42          BETA=(RHO/RHO_OLD)*(ALPHA_OLD/OM_OLD)
43          P=R_OLD+BETA*(P_OLD-OM_OLD*V_OLD)
44       END IF
45    !-------------------------------------------------
46       CALL LUSOL    (N,NNZERO,P,PHAT,LUVAL,JA_AUX,IA_AUX,UPTR)
47    !-------------------------------------------------
48    !Stores the matrix in compressed spars row and multiplies by
49    !vector
50    !-------------------------------------------------
51    !CALL CSR_MAT_V_PRODUCT     (N,NNZERO,AA_AUX,JA_AUX,IA_AUX,PHAT,V)
52    !-------------------------------------------------
53    ! Stores the matrix in modified spars row and multiplies by
54    ! vector
55    !-------------------------------------------------
56       CALL MSR_MAT_V_PRODUCT (TL,N,SA,IJA,PHAT,V)
57    !-------------------------------------------------
58    !Stores the matrix in Ellpack/Itpack format and multiplies by
59    !vector
60    !-------------------------------------------------
61    !CALL ELLPACK_ITPACK_MAT_V_PRODUCT (N,PHAT,V,N,IDIAG,COEF,JCOEF)
62    !-------------------------------------------------
63    !Stores the matrix in Diagonal format and multiplies by
64    !vector
65    !-------------------------------------------------
66    !CALL DIAGONAL_MAT_V_PRODUCT (N,PHAT,V,DIAG,N,IDIAG,IOFF)
67    !-------------------------------------------------
68       ALPHA=RHO/DOT_PRODUCT(RS,V)
69       SR=R_OLD-ALPHA*V
70    !-------------------------------------------------
71       CALL LUSOL    (N,NNZERO,SR,SRHAT,LUVAL,JA_AUX,IA_AUX,UPTR)
72    !-------------------------------------------------
73    !Stores the matrix in compressed spars row and multiplies by
74    !vector
75    !-------------------------------------------------
76    !CALL CSR_MAT_V_PRODUCT     (N,NNZERO,AA_AUX,JA_AUX,IA_AUX,SRHAT,T)
77    !-------------------------------------------------
78    ! Stores the matrix in modified spars row and multiplies by
79    ! vector
80    !-------------------------------------------------
81       CALL MSR_MAT_V_PRODUCT (TL,N,SA,IJA,SRHAT,T)
82    !-------------------------------------------------
83    !Stores the matrix in Ellpack/Itpack format and multiplies by
84    !vector
85    !-------------------------------------------------
86    !CALL ELLPACK_ITPACK_MAT_V_PRODUCT (N,SRHAT,T,N,IDIAG,COEF,JCOEF)
87    !-------------------------------------------------
88    !Stores the matrix in Diagonal format and multiplies by
89    !vector
90    !-------------------------------------------------
91    !CALL DIAGONAL_MAT_V_PRODUCT (N,SRHAT,T,DIAG,N,IDIAG,IOFF)
92    !-------------------------------------------------
93       OM=DOT_PRODUCT(T,SR)/DOT_PRODUCT(T,T)
94       X= X_OLD+ALPHA*PHAT+OM*SRHAT
95       R=SR-OM*T
96       X_OLD=X
97       R_OLD=R
98       RHO_OLD=RHO
99       P_OLD=P
100      OM_OLD=OM
101      V_OLD=V
102      ALPHA_OLD=ALPHA
103   END DO
104   !-------------------------------------------------
105   END SUBROUTINE ILU0_PBCGSTAB_INNER
```

Appendix E

Main Program

The main program is the kernel of the code and controls the inputs, processes and outputs. To construct the system of equations one can open a matrix data file with extension .MTX where its data is arranged in coordinate format. A special matrix is determined with integer flag JOB1 and integer flag JOB2 denotes its symmetry or asymmetry. The subroutine MATRIX_PATTERN is called to save the matrix shape; see Appendix A.4 for details. To construct the right-hand side vector of the equations system, a vector containing the unit entries is multiplied by the selected matrix in CSR format with calling the subroutine CSR_MAT_V_PRODUCT; see Section 4.1 for details. The prerequisite of this multiplication is the coordinate to CSR format conversion via the subroutine COOCSR; for more details see Appendix A.1.1.

```
1  !************************************************
2  !                Main Program
3  !************************************************
4  PROGRAM MAIN
5  !--------------------------------------------------
6  IMPLICIT NONE
7  INTEGER,ALLOCATABLE   :: REV_IR(:),REV_JC(:),IAO(:),IAOO(:),IR(:),JC(:)
8  INTEGER,ALLOCATABLE   :: IRR(:),JCC(:),IVAL(:),JVAL(:),JAO(:),JAOO(:)
9  REAL(8),ALLOCATABLE   :: X(:),B(:),SOL(:),AA(:),AO(:),AOO(:),VALU(:),
        REV_A(:),A(:)
10 INTEGER               :: JOB1,JOB2,I,J,K,TNNZ,NNZ,N
11 !--------------------------------------------------------------------
12 !NAME OF MATRICES | JOB1  |  TYPE |  TNNZ  |  N   |  NNZ
13 !--------------------------------------------------------------------
14 !PLAT1919         |   1   |  SI   | 32399  | 1919 | 17159
15 !PLAT362          |   2   |  SI   |  5786  |  362 |  3074
16 !GR3030           |   3   |  SPD  |  7744  |  900 |  4322
17 !BCSSTK22         |   4   |  SI   |   696  |  138 |   417
18 !NOS4             |   5   |  SPD  |   594  |  100 |   347
19 !HOR131           |   6   |  NS   |  4710  |  434 |  --
20 !NOS6             |   7   |  SPD  |  3255  |  675 |  1965
21 !NOS7             |   8   |  SPD  |  4617  |  729 |  2673
22 !ORSIRR1          |   9   |  NS   |  6858  | 1030 |  --
23 !SHERMAN4         |  10   |  NS   |  3786  | 1104 |  --
24 !ORSIRR2          |  11   |  NS   |  5970  |  886 |  --
```

Krylov Subspace Methods with Application in Incompressible Fluid Flow Solvers, First Edition.
Iman Farahbakhsh.
© 2020 John Wiley & Sons Ltd. Published 2020 by John Wiley & Sons Ltd.
Companion Website: www.wiley.com/go/Farahbakhs/KrylovSubspaceMethods

```
25  !ORSREG1                    |    12   |   NS   |  14133  |  2205  |  --
26  !---------------------------------------------------------------
27  !   SI:  SYMMETRIC INDEFINITE
28  !   SPD: SYMMETRIC POSITIVE DEFINITE
29  !   NS:  NONSYMMETRIC
30  !   JOB2=0 FOR SYMMETRIC MATRICES
31  !---------------------------------------------------------------
32  JOB1=3
33  JOB2=0
34  !---------------------------------------------------------------
35  IF (JOB1.EQ.1) THEN
36  OPEN(UNIT=JOB1,FILE='plat1919.MTX')
37  TNNZ=32399
38  NNZ=17159
39  N=1919
40  ELSE IF (JOB1.EQ.2) THEN
41  OPEN(UNIT=JOB1,FILE='plat362.MTX')
42  TNNZ=5786
43  NNZ=3074
44  N=362
45  ELSE IF (JOB1.EQ.3) THEN
46  OPEN(UNIT=JOB1,FILE='gr_30_30.MTX')
47  TNNZ=7744
48  NNZ=4322
49  N=900
50  ELSE IF (JOB1.EQ.4) THEN
51  OPEN(UNIT=JOB1,FILE='bcsstk22.MTX')
52  TNNZ=696
53  NNZ=417
54  N=138
55  ELSE IF (JOB1.EQ.5) THEN
56  OPEN(UNIT=JOB1,FILE='nos4.MTX')
57  TNNZ=594
58  NNZ=347
59  N=100
60  ELSE IF (JOB1.EQ.6) THEN
61  OPEN(UNIT=JOB1,FILE='hor__131.MTX')
62  TNNZ=4710
63  N=434
64  ELSE IF (JOB1.EQ.7) THEN
65  OPEN(UNIT=JOB1,FILE='nos6.MTX')
66  TNNZ=3255
67  NNZ=1965
68  N=675
69  ELSE IF (JOB1.EQ.8) THEN
70  OPEN(UNIT=JOB1,FILE='nos7.MTX')
71  TNNZ=4617
72  NNZ=2673
73  N=729
74  ELSE IF (JOB1.EQ.9) THEN
75  OPEN(UNIT=JOB1,FILE='orsirr_1.MTX')
76  TNNZ=6858
77  N=1030
78  ELSE IF (JOB1.EQ.10) THEN
79  OPEN(UNIT=JOB1,FILE='sherman4.MTX')
80  TNNZ=3786
81  N=1104
82  ELSE IF (JOB1.EQ.11) THEN
83  OPEN(UNIT=JOB1,FILE='orsirr_2.MTX')
84  TNNZ=5970
85  N=886
86  ELSE IF (JOB1.EQ.12) THEN
87  OPEN(UNIT=JOB1,FILE='orsreg_1.MTX')
88  TNNZ=14133
89  N=2205
90  END IF
```

```
 91   !-------------------------------------------------
 92   ALLOCATE (IRR(TNNZ),JCC(TNNZ),IVAL(TNNZ),JVAL(TNNZ),JAO(TNNZ),
          JAOO(TNNZ))
 93   ALLOCATE (IAO(N+1),IAOO(N+1),AA(TNNZ),AO(TNNZ),AOO(TNNZ),VALU(TNNZ),
          X(N),B(N),SOL(N))
 94   IF (JOB2.EQ.0) THEN
 95      ALLOCATE (IR(NNZ),JC(NNZ),REV_IR(TNNZ-NNZ),REV_JC(TNNZ-NNZ),A(NNZ),
          REV_A(TNNZ-NNZ))
 96      DO I=1,NNZ
 97         READ(JOB1,*) IR(I),JC(I),A(I)  !Read the content of the file
 98      END DO
 99   !-------------------------------------------------
100      K=0
101      DO J=1,NNZ
102         IF (IR(J).NE.JC(J)) THEN
103            K=K+1
104            REV_IR(K)=JC(J)
105            REV_JC(K)=IR(J)
106            REV_A(K)=A(J)
107         END IF
108      END DO
109   !-------------------------------------------------
110      DO I=1,TNNZ
111         IF (I.LE.NNZ) THEN
112            IRR(1:NNZ)=IR(1:NNZ)
113            JCC(1:NNZ)=JC(1:NNZ)
114            AA(1:NNZ)=A(1:NNZ)
115         ELSE
116            IRR(NNZ+1:TNNZ)=REV_IR(1:TNNZ-NNZ)
117            JCC(NNZ+1:TNNZ)=REV_JC(1:TNNZ-NNZ)
118            AA(NNZ+1:TNNZ)=REV_A(1:TNNZ-NNZ)
119         END IF
120      END DO
121      CALL MATRIX_PATTERN (JOB1,TNNZ,N,IRR,JCC)
122      CALL COOCSR (N,TNNZ,AA,IRR,JCC,AO,JAO,IAO)
123      X=1.D0
124      CALL CSR_MAT_V_PRODUCT (N,TNNZ,AO,JAO,IAO,X,B)
125      CALL SOLVERS (N,TNNZ,B,AO,JAO,IAO,SOL)
126   !-------------------------------------------------
127   ELSE
128      DO I=1,TNNZ
129         READ(JOB1,*) IVAL(I),JVAL(I),VALU(I)  !Read the content of the
             file
130      END DO
131      CALL MATRIX_PATTERN (JOB1,TNNZ,N,IVAL,JVAL)
132      CALL COOCSR (N,TNNZ,VALU,IVAL,JVAL,AOO,JAOO,IAOO)
133      X=1.D0
134      CALL CSR_MAT_V_PRODUCT (N,TNNZ,AOO,JAOO,IAOO,X,B)
135      CALL SOLVERS (N,TNNZ,B,AOO,JAOO,IAOO,SOL)
136   END IF
137   CALL WRITE_RESULT (JOB1,N,X,SOL)
138   !-------------------------------------------------
139   END PROGRAM MAIN
```

In the PROGRAM MAIN, solvers can be called via subroutine SOLVERS. On entry, MD and NNZERO denote the matrix dimension and number of non-zero elements. On entry list, we have RHS as the right-hand side vector and the matrix in CSR format which is presented here as AA_AUX, JA_AUX and IA_AUX. On return, SOL is the solution vector of length MD. In this subroutine, the user can activate the CALL command of a Krylov subspace method, with uncommenting the corresponding line. One can find the calling command of nine solvers subroutines of Appendices B.1 to B.6 and Appendices C.1 to C.3 from line 22 to

30, where as mentioned in Appendix B.6, the subroutine GMRES∗ contains seven subroutines of Appendices D.1 to D.7, which act as the preconditioner.

```
 1 !*************************************************
 2 !                    Solvers
 3 !*************************************************
 4 SUBROUTINE SOLVERS (MD,NNZERO,RHS,AA_AUX,JA_AUX,IA_AUX,SOL)
 5 !-----------------------------------------------
 6 IMPLICIT NONE
 7 INTEGER                             :: NNZERO,MD
 8 REAL(8)                             :: BEGIN_TIME,END_TIME
 9 REAL(8),DIMENSION(MD)               :: RHS
10 REAL(8),DIMENSION(MD)               :: X0,R0,SOL
11 REAL(8),DIMENSION(NNZERO)           :: AA_AUX
12 INTEGER,DIMENSION(NNZERO)           :: JA_AUX
13 INTEGER,DIMENSION(MD+1)             :: IA_AUX
14 !-----------------------------------------------
15 CALL INITIAL_GUESS (MD,X0)
16 !-----------------------------------------------
17 CALL INITIAL_RESIDUAL (MD,NNZERO,AA_AUX,JA_AUX,IA_AUX,RHS,X0,R0)
18 !-----------------------------------------------
19 !Krylov subspaces methods
20 !-----------------------------------------------
21 CALL CPU_TIME (BEGIN_TIME)
22 !-----------------------------------------------
23 !CALL CG             (MD,NNZERO,AA_AUX,JA_AUX,IA_AUX,X0,R0,SOL)
24 !CALL BCG            (MD,NNZERO,AA_AUX,JA_AUX,IA_AUX,X0,R0,SOL)
25 !CALL CGS            (MD,NNZERO,AA_AUX,JA_AUX,IA_AUX,X0,R0,SOL)
26 !CALL BCGSTAB        (MD,NNZERO,AA_AUX,JA_AUX,IA_AUX,X0,R0,SOL)
27 !CALL CR             (MD,NNZERO,AA_AUX,JA_AUX,IA_AUX,X0,R0,SOL)
28 !CALL ILU0_PCG       (MD,NNZERO,AA_AUX,JA_AUX,IA_AUX,X0,R0,SOL)
29 !CALL ILU0_PCGS      (MD,NNZERO,AA_AUX,JA_AUX,IA_AUX,X0,R0,SOL)
30 !CALL ILU0_PBCGSTAB  (MD,NNZERO,AA_AUX,JA_AUX,IA_AUX,X0,R0,SOL)
31 CALL GMRES_STAR     (MD,NNZERO,AA_AUX,JA_AUX,IA_AUX,X0,R0,SOL)
32 !-----------------------------------------------
33 CALL CPU_TIME (END_TIME)
34 !OPEN(88888,FILE='cput.txt',STATUS='unknown')
35 !WRITE(88888,*) END_TIME-BEGIN_TIME
36 PRINT*,'Total spent time in Krylov solver is',END_TIME-BEGIN_TIME,'sec'
37 !-----------------------------------------------
38 END SUBROUTINE SOLVERS
```

The subroutines INITIAL_GUESS and INITIAL_RESIDUAL, as their names imply, are called to produce the initial guess and consequently, the initial residual for beginning the solution process. In the subroutine INITIAL_GUESS, the vector X0 is simply set to zero as the initial guess. The vector X0 is transfered on the entry list of the subroutine INITIAL_RESIDUAL to produce the initial residual R0. The rest of the entries list is the same as the entries list of the subroutine SOLVERS.

```
 1 !*************************************************
 2 !                Initial Guess
 3 !*************************************************
 4 SUBROUTINE INITIAL_GUESS (MD,X0)
 5 !-----------------------------------------------
 6 IMPLICIT NONE
 7 INTEGER                  :: MD
 8 REAL(8),DIMENSION(MD)    :: X0
 9 !-----------------------------------------------
10 X0=0.D0
11 !-----------------------------------------------
12 END SUBROUTINE INITIAL_GUESS
```

```
1  !******************************************************
2  !                  Initial Residual
3  !******************************************************
4  SUBROUTINE INITIAL_RESIDUAL (MD,NNZERO,AA_AUX,JA_AUX,IA_AUX,RHS,X0,R0)
5  !----------------------------------------------------
6  IMPLICIT NONE
7  INTEGER                        :: MD,NNZERO
8  REAL(8),DIMENSION(MD)          :: RHS,X0,R0,Y0
9  REAL(8),DIMENSION(NNZERO)      :: AA_AUX
10 INTEGER,DIMENSION(NNZERO)      :: JA_AUX
11 INTEGER,DIMENSION(MD+1)        :: IA_AUX
12 !----------------------------------------------------
13 CALL CSR_MAT_V_PRODUCT (MD,NNZERO,AA_AUX,JA_AUX,IA_AUX,X0,Y0)
14 !----------------------------------------------------
15 R0=RHS-Y0
16 !----------------------------------------------------
17 END SUBROUTINE INITIAL_RESIDUAL
```

The subroutine WRITE_RESULT can write the solution vector or exact error of the corresponding system of equations with the coefficients matrix of the flag JOB1. The results are saved with . PLT extension.

```
1  !******************************************************
2  !                  Write Result
3  !******************************************************
4  SUBROUTINE WRITE_RESULT (JOB1,N,X,SOL)
5  !----------------------------------------------------
6  IMPLICIT NONE
7  INTEGER                  :: I,N,JOB1
8  REAL(8),DIMENSION(N)     :: X,SOL
9  !----------------------------------------------------
10 IF (JOB1.EQ.1) THEN
11        OPEN(UNIT=JOB1+1000,FILE='error_plat1919.PLT')
12 ELSE IF (JOB1.EQ.2) THEN
13        OPEN(UNIT=JOB1+1000,FILE='error_plat362.PLT')
14 ELSE IF (JOB1.EQ.3) THEN
15        OPEN(UNIT=JOB1+1000,FILE='error_gr_30_30.PLT')
16 ELSE IF (JOB1.EQ.4) THEN
17        OPEN(UNIT=JOB1+1000,FILE='error_bcsstk22.PLT')
18 ELSE IF (JOB1.EQ.5) THEN
19        OPEN(UNIT=JOB1+1000,FILE='error_nos4.PLT')
20 ELSE IF (JOB1.EQ.6) THEN
21        OPEN(UNIT=JOB1+1000,FILE='error_hor__131.PLT')
22 ELSE IF (JOB1.EQ.7) THEN
23        OPEN(UNIT=JOB1+1000,FILE='error_nos6.PLT')
24 ELSE IF (JOB1.EQ.8) THEN
25        OPEN(UNIT=JOB1+1000,FILE='error_nos7.PLT')
26 ELSE IF (JOB1.EQ.9) THEN
27        OPEN(UNIT=JOB1+1000,FILE='error_orsirr_1.PLT')
28 ELSE IF (JOB1.EQ.10) THEN
29        OPEN(UNIT=JOB1+1000,FILE='error_sherman4.PLT')
30 ELSE IF (JOB1.EQ.11) THEN
31        OPEN(UNIT=JOB1+1000,FILE='error_orsirr_2.PLT')
32 ELSE IF (JOB1.EQ.12) THEN
33        OPEN(UNIT=JOB1+1000,FILE='error_orsreg_1.PLT')
34 END IF
35 DO I=1,N
36 WRITE (JOB1+1000,*) I,ABS(SOL(I)-X(I))
37 END DO
38 !----------------------------------------------------
39 END SUBROUTINE WRITE_RESULT
```

Appendix F

Steepest Descent Method

```
1   ! This program is developed for extracting the
2   ! convergence trace of steepest descent method
3   ! when solves a linear system of equations, Ax=b,
4   ! where A is a symmetric positive definite matrix.
5   !--------------------------------------------------
6   MODULE SD
7   IMPLICIT NONE
8   REAL(8),PARAMETER              ::  EPS    = 1.D-10
9   END MODULE
10  !--------------------------------------------------
11  PROGRAM MAIN
12  USE SD, ONLY                     :  EPS
13  IMPLICIT NONE
14  INTEGER                          ::  I,J,N
15  REAL(8),ALLOCATABLE              ::  A(:,:),B(:),X(:)
16  !--------------------------------------------------
17  PRINT*,"This program is developed for extracting the convergence trace
        of&
18          & steepest descent method when solves a linear system of
        equations,&
19          & Ax=b, where A is a symmetric positive definite matrix."
20  PRINT*,""
21  PRINT*, "Please insert the order of matrix..."
22  PRINT*,""
23  READ*, N
24  PRINT*,""
25  ALLOCATE(A(N,N),B(N),X(N))
26  PRINT*, "Please insert the elements of a symmetric positive definite
        matrix&
27          &, row by row. After inserting each row press enter  to move
        the&
28          & cursor to the next line. It should be noted that elements
        must&
29          & be separated with at least one space. "
30  PRINT*,""
```

Krylov Subspace Methods with Application in Incompressible Fluid Flow Solvers, First Edition.
Iman Farahbakhsh.
© 2020 John Wiley & Sons Ltd. Published 2020 by John Wiley & Sons Ltd.
Companion Website: www.wiley.com/go/Farahbakhs/KrylovSubspaceMethods

```
31 DO I=1,N
32 READ*, (A(I,J),J=1,N)
33 END DO
34 PRINT*, ""
35 PRINT*, "Please insert the elements of right hand side vector.&
36        & Please press enter after inserting each element to move the
         cursor to the next line."
37 PRINT*,""
38 DO I=1,N
39 READ*, B(I)
40 END DO
41 CALL STEEPEST_DESCENT_METHOD (N,A,B,X)
42 !-----------------------------------------------
43 END PROGRAM
44 !**************************************************
45 !        SUCCESSIVE SOLUTION SUBROUTINE
46 !**************************************************
47 SUBROUTINE SUCCESSIVE_SOL (N,X,ITER)
48 !-----------------------------------------------
49 IMPLICIT NONE
50 INTEGER                           :: ITER,COUNTER,N
51 CHARACTER*10                      :: EXT
52 CHARACTER*4                       :: FN1
53 CHARACTER*25                      :: FNAME
54 REAL(8),DIMENSION(N)              :: X
55 !-----------------------------------------------
56 FN1='VECT'
57 COUNTER=0
58 !-----------------------------------------------
59 COUNTER=COUNTER+10
60 WRITE(EXT,'(I7)') ITER
61 FNAME=FN1//EXT//'.DAT'
62 !-----------------------------------------------
63 OPEN(COUNTER,FILE=FNAME,POSITION='REWIND')
64 !WRITE(COUNTER,*)'VARIABLES= "X"'
65 !WRITE(COUNTER,*)'ZONE,F=POINT'
66 WRITE(COUNTER,*) X
67 CLOSE(COUNTER)
68 !-----------------------------------------------
69 WRITE(*,*) '========================'
70 WRITE(*,*) 'PRINTING ON ',FNAME
71 WRITE(*,*) '========================'
72 !-----------------------------------------------
73 END SUBROUTINE SUCCESSIVE_SOL
74 !**************************************************
75 !        STEEPEST DESCENT METHOD SUBROUTINE
76 !**************************************************
77 SUBROUTINE STEEPEST_DESCENT_METHOD (N,A,B,X)
78 !-----------------------------------------------
79 USE SD
80 IMPLICIT NONE
81 INTEGER                    :: ITER,I,N
82 REAL(8)                    :: LAMBDA,RNORM,S,ARR
83 REAL(8),DIMENSION(N,N)     :: A
84 REAL(8),DIMENSION(N)       :: B,X,RSDL,AR,AX
85 !-----------------------------------------------
86 X=0.D0
87 RNORM=1.D0
88 ITER=0
89 DO WHILE (RNORM.GT.EPS)
90 ITER=ITER+1
91 PRINT*,ITER
92 AX=MATMUL(A,X)
93 RSDL=B-AX
94 AR=MATMUL(A,RSDL)
95 ARR=DOT_PRODUCT(AR,RSDL)
96 S=0.D0
```

```
 97  DO I=1,N
 98  S=S+RSDL(I)*RSDL(I)
 99  END DO
100  RNORM=SQRT(S)
101  LAMBDA=S/ARR
102  X=X+LAMBDA*RSDL
103  CALL SUCCESSIVE_SOL (N,X,ITER)
104  PRINT*,RNORM
105  END DO
106  !PRINT*,X
107  !-------------------------------------------------
108  END SUBROUTINE STEEPEST_DESCENT_METHOD
```

Appendix G

Vorticity-Stream Function Formulation of Navier–Stokes Equation

```
 1  !**************************************************
 2  !               Module Parameters
 3  !**************************************************
 4  MODULE PARAM
 5  !-------------------------------------------------
 6  IMPLICIT NONE
 7  INTEGER,PARAMETER                    :: IM=101
 8  INTEGER,PARAMETER                    :: JM=101
 9  INTEGER,PARAMETER                    :: NODE_NUM=IM*JM
10  INTEGER,PARAMETER                    :: ITMAX=1000
11  REAL(8),PARAMETER                    :: LX=1.D0
12  REAL(8),PARAMETER                    :: LY=1.D0
13  REAL(8),PARAMETER                    :: DXI=LX/(IM-1)
14  REAL(8),PARAMETER                    :: DETA=LY/(JM-1)
15  REAL(8),PARAMETER                    :: TOL=1.D-10
16  INTEGER,PARAMETER                    :: NSTEP=300000
17  INTEGER,PARAMETER                    :: NPRINT=100
18  INTEGER,PARAMETER                    :: KSTART=0
19  REAL(8),PARAMETER                    :: DT=1.D-3
20  REAL(8),PARAMETER                    :: RE=1000.D0
21  END MODULE PARAM
22  !**************************************************
23  !                Main Program
24  !**************************************************
25  PROGRAM MAIN
26  !-------------------------------------------------
27  USE PARAM, ONLY            : IM,JM,NODE_NUM,DT,NSTEP,KSTART,
        NPRINT
28  IMPLICIT NONE
29  INTEGER                              :: K,MD,NNZERO
30  INTEGER,DIMENSION((IM-2)*(JM-2))     :: RHS_POINTER
31  REAL(8),DIMENSION((IM-2)*(JM-2))     :: RHS
32  REAL(8),DIMENSION(IM,JM)             :: XI,ETA,U,V,SI,OMEGA,RHS_VCE
33  REAL(8),DIMENSION(NODE_NUM)          :: XUS,YUS,F,COEFF_C,COEFF_B,COEFF_L,
        COEFF_R,COEFF_T
34  LOGICAL,DIMENSION(NODE_NUM)          :: NODE_TYPE
35  INTEGER,DIMENSION(NODE_NUM)          :: L_NEIGHBOR,R_NEIGHBOR,U_NEIGHBOR,
        D_NEIGHBOR
36  REAL(8),ALLOCATABLE                  :: AA_AUX(:)
37  INTEGER,ALLOCATABLE                  :: JA_AUX(:),IA_AUX(:),JA_AUX_POINTER
        (:)
38  !-------------------------------------------------
39  CALL STRUCT_GRID            (XI,ETA)
40  CALL UNSTRUCT_GRID          (XI,ETA,XUS,YUS,NODE_TYPE)
```

Krylov Subspace Methods with Application in Incompressible Fluid Flow Solvers, First Edition.
Iman Farahbakhsh.
© 2020 John Wiley & Sons Ltd. Published 2020 by John Wiley & Sons Ltd.
Companion Website: www.wiley.com/go/Farahbakhs/KrylovSubspaceMethods

```
41 CALL NEIGHBORS                  (NODE_TYPE,L_NEIGHBOR,R_NEIGHBOR,U_NEIGHBOR,
      D_NEIGHBOR)
42 !-----------------------------------------------------
43 CALL RHS_CONSTITUTION (F,COEFF_C,COEFF_B,COEFF_L,COEFF_R,COEFF_T,
      L_NEIGHBOR,R_NEIGHBOR&
44 ,U_NEIGHBOR,D_NEIGHBOR,NODE_TYPE,SI,RHS_POINTER,RHS,MD,NNZERO)
45 !-----------------------------------------------------
46 ALLOCATE (AA_AUX(NNZERO),JA_AUX_POINTER(NNZERO),JA_AUX(NNZERO),IA_AUX(
      MD+1))
47 !-----------------------------------------------------
48 CALL MATRIX_CONSTITUTION (NNZERO,MD,COEFF_C,COEFF_B,COEFF_L,COEFF_R,
      COEFF_T&
49 ,NODE_TYPE,D_NEIGHBOR,L_NEIGHBOR,R_NEIGHBOR,U_NEIGHBOR,AA_AUX,
      JA_AUX_POINTER,IA_AUX,JA_AUX)
50 !-----------------------------------------------------
51 DO K=1,NSTEP
52 CALL IBC                        (NODE_TYPE,SI,OMEGA,F)
53 CALL RHS_CONSTITUTION (F,COEFF_C,COEFF_B,COEFF_L,COEFF_R,COEFF_T,
      L_NEIGHBOR,R_NEIGHBOR&
54 ,U_NEIGHBOR,D_NEIGHBOR,NODE_TYPE,SI,RHS_POINTER,RHS,MD,NNZERO)
55 CALL SOLVERS                    (MD,NNZERO,RHS,RHS_POINTER,AA_AUX,JA_AUX,
      IA_AUX,SI)        ! Elliptic part
56 CALL OMEGA_BC (SI,OMEGA)
57 CALL U_V (SI,U,V)
58 CALL DERIVE (OMEGA,U,V,RHS_VCE)
59 CALL RK4 (OMEGA,U,V,RHS_VCE)                    ! Fourth-order Runge-Kutta
60 PRINT*,K
61 IF(((K/NPRINT)*NPRINT)==K)  CALL WRITE_TOTAL_SOL (XI,ETA,U,V,SI,OMEGA,K
      )
62 END DO
63 !-----------------------------------------------------
64 END PROGRAM MAIN
65 !********************************************************
66 !       Vorticity Transport Equation RHS
67 !********************************************************
68 SUBROUTINE DERIVE (OMEGA,U,V,RHS_VCE)
69 !-----------------------------------------------------
70 USE PARAM    ,ONLY          : IM,JM,DXI,DETA,RE
71 IMPLICIT NONE
72 INTEGER                ::I,J
73 REAL(8),DIMENSION(IM,JM) ::U,V,OMEGA,RHS_VCE
74 !-----------------------------------------------------
75 DO I=2,IM-1
76 DO J=2,JM-1
77   RHS_VCE(I,J)=(OMEGA(I+1,J)-2*OMEGA(I,J)+OMEGA(I-1,J))/(RE*DXI*DXI)&
78               +(OMEGA(I,J+1)-2*OMEGA(I,J)+OMEGA(I,J-1))/(RE*DETA*DETA)
      &
79               -0.5*U(I,J)*(OMEGA(I+1,J)-OMEGA(I-1,J))/DXI&
80               -0.5*V(I,J)*(OMEGA(I,J+1)-OMEGA(I,J-1))/DETA
81 END DO
82 END DO
83 !-----------------------------------------------------
84 END SUBROUTINE DERIVE
85 !********************************************************
86 !           Velocity Field BC
87 !********************************************************
88 SUBROUTINE U_V_BC (U,V)
89 !-----------------------------------------------------
90 USE PARAM, ONLY            : IM,JM
91 IMPLICIT NONE
92 REAL(8),DIMENSION(IM,JM)   :: U,V
93 !-----------------------------------------------------
94 U(:,JM)=1.D0
95 V(:,JM)=0.D0
96 U(:,1)=0.D0
97 V(:,1)=0.D0
98 U(1,:)=0.D0
```

```
 99  V(1,:)=0.D0
100  U(IM,:)=0.D0
101  V(IM,:)=0.D0
102  !----------------------------------------------------
103  END SUBROUTINE U_V_BC
104  !***************************************************
105  !                 Vorticity BC
106  !***************************************************
107  SUBROUTINE OMEGA_BC (SI,OMEGA)
108  !----------------------------------------------------
109  USE PARAM ,ONLY    : IM,JM,DETA,DXI
110  IMPLICIT NONE
111  INTEGER                    :: I,J
112  REAL(8),DIMENSION(IM,JM)   :: SI,OMEGA
113  !----------------------------------------------------
114  DO J=1,JM
115     OMEGA(1,J)=2*(SI(1 ,J)-SI(2   ,J))/(DXI)**2
116     OMEGA(IM,J)=2*(SI(IM,J)-SI(IM-1,J))/(DXI)**2
117  END DO
118  DO I=1,IM
119     OMEGA(I,JM)=2*(SI(I,JM)-SI(I,JM-1))/(DETA)**2-2.D0/(DETA)
120     OMEGA(I,1)=2*(SI(I,1)-SI(I,2))/(DETA)**2
121  END DO
122  !----------------------------------------------------
123  END SUBROUTINE OMEGA_BC
124  !***************************************************
125  !           Velocity Field Derivation
126  !***************************************************
127  SUBROUTINE U_V (SI,U,V)
128  !----------------------------------------------------
129  USE PARAM ,ONLY              : IM,JM,DXI,DETA
130  IMPLICIT NONE
131  INTEGER                    :: I,J
132  REAL(8),DIMENSION(IM,JM)   :: U,V,SI
133  !----------------------------------------------------
134  DO I=2,IM-1
135     DO J=2,JM-1
136        U(I,J)=+(SI(I,J+1)-SI(I,J-1))/(2*DETA)
137        V(I,J)=-(SI(I+1,J)-SI(I-1,J))/(2*DXI)
138     END DO
139  END DO
140  !----------------------------------------------------
141  ENDSUBROUTINE U_V
142  !***************************************************
143  !         Fourth Order Runge-Kutta
144  !***************************************************
145  SUBROUTINE RK4 (OMEGA,U,V,RHS_VCE)
146  !----------------------------------------------------
147  USE PARAM   ,ONLY          : IM,JM,DT
148  IMPLICIT NONE
149  REAL(8),DIMENSION(IM,JM)      :: OMEGA,OMEGA0,U,V,RHS_VCE,K1,K2,K3,K4
150  !----------------------------------------------------
151  OMEGA0=OMEGA
152  CALL DERIVE (OMEGA,U,V,RHS_VCE)
153  K1=RHS_VCE
154  OMEGA=OMEGA0+0.5*DT*K1
155  CALL DERIVE (OMEGA,U,V,RHS_VCE)
156  K2=RHS_VCE
157  OMEGA=OMEGA0+0.5*DT*K2
158  CALL DERIVE (OMEGA,U,V,RHS_VCE)
159  K3=RHS_VCE
160  OMEGA=OMEGA0+DT*K3
161  CALL DERIVE (OMEGA,U,V,RHS_VCE)
162  K4=RHS_VCE
163  OMEGA=OMEGA0+DT*(K1+2*K2+2*K3+K4)/6
164  !----------------------------------------------------
165  END SUBROUTINE RK4
```

```fortran
166 !****************************************************
167 !            Write Total Solution
168 !****************************************************
169 SUBROUTINE WRITE_TOTAL_SOL (XI,ETA,U,V,SI,OMEGA,K)
170 !--------------------------------------------------
171 USE PARAM    ,ONLY    : IM,JM,KSTART
172 IMPLICIT NONE
173 INTEGER                              :: I,J,K,COUNTER
174 REAL(8),DIMENSION(IM,JM)             :: U,V,XI,ETA
175 REAL(8),DIMENSION(IM,JM),INTENT(IN)  :: OMEGA,SI
176 CHARACTER*7                          :: EXT
177 CHARACTER*3                          :: FN1
178 CHARACTER*16                         :: FNAME
179 !--------------------------------------------------
180 CALL U_V_BC (U,V)
181 FN1='FLD'
182 !--------------------------------------------------
183 COUNTER=COUNTER+10
184 WRITE(EXT,'(I5)') K+KSTART
185 FNAME=FN1//EXT//'.DAT'
186 !--------------------------------------------------
187 OPEN(COUNTER,FILE=FNAME,POSITION='REWIND')
188 WRITE(COUNTER,*)'VARIABLES="XI","ETA","U","V","SI","OMEGA"'
189 WRITE(COUNTER,*)'ZONE, J=',JM,',I=',IM,',F=POINT'
190 DO J=1,JM
191    DO I=1,IM
192       WRITE(COUNTER,*) XI(I,J),ETA(I,J),U(I,J),V(I,J),SI(I,J),OMEGA(I,J
       )
193 !100 FORMAT (6F15.8)
194    END DO
195 END DO
196 CLOSE(COUNTER)
197 !--------------------------------------------------
198 WRITE(*,*) '======================='
199 WRITE(*,*) 'PRINTING ON ',FNAME
200 WRITE(*,*) '======================='
201 !--------------------------------------------------
202 END SUBROUTINE WRITE_TOTAL_SOL
203 !****************************************************
204 !               Unstructured Grid
205 !****************************************************
206 !UNSTRUCT_GRID Subroutine
207 !This subroutine which has been written
208 !by Iman Farahbakhsh on Jan 2010 enable
209 !us to extract the node number for each
210 !grid point with i and j indices. The
211 !type of each node (boundary-node or middle-node)
212 !is also determined as a logical parameter.
213 !
214 !on entry
215 !===========
216 !X,Y       The positions which are stored by i and j indices
217 !
218 !on exit
219 !===========
220 !XUS,YUS   The positions which are stored by node index
221 !
222 !NODE_TYPE A logical array which shows that a node is boundary or not
223 !
224 !Last change: Jan 17 2010
225 !--------------------------------------------------
226 SUBROUTINE UNSTRUCT_GRID (X,Y,XUS,YUS,NODE_TYPE)
227 !--------------------------------------------------
228 USE PARAM, ONLY            : IM,JM,NODE_NUM
229 IMPLICIT NONE
230 INTEGER                     :: INODE,I,J
231 REAL(8),DIMENSION(IM,JM)    :: X,Y
```

```fortran
232   REAL(8),DIMENSION(NODE_NUM)        :: XUS,YUS
233   LOGICAL,DIMENSION(NODE_NUM)        :: NODE_TYPE
234   !-------------------------------------------------
235   DO J=1,JM
236      DO I=1,IM
237         IF (J.EQ.1) THEN
238            INODE=I
239            NODE_TYPE(INODE)=.TRUE.
240          XUS(INODE)=X(I,J)
241          YUS(INODE)=Y(I,J)
242          ELSEIF (J.EQ.JM) THEN
243            INODE=(JM-1)*IM+I
244            NODE_TYPE(INODE)=.TRUE.
245             XUS(INODE)=X(I,J)
246          YUS(INODE)=Y(I,J)
247          ELSE
248            INODE=(J-1)*IM+I
249             XUS(INODE)=X(I,J)
250             YUS(INODE)=Y(I,J)
251         IF (I.EQ.1.OR.I.EQ.IM) THEN
252            NODE_TYPE(INODE)=.TRUE.
253             XUS(INODE)=X(I,J)
254          YUS(INODE)=Y(I,J)
255         END IF
256         END IF
257      END DO
258   END DO
259   !-------------------------------------------------
260   WRITE(*,*) '******* Nodes number were assigned *******'
261   !-------------------------------------------------
262   END SUBROUTINE UNSTRUCT_GRID
263   !*************************************************
264   !                 Structured Grid
265   !*************************************************
266   !STRUCT_GRID Subroutine
267   !This subroutine which has been written by Iman Farahbakhsh on Oct 2009 enable
268   !us to generate a uniform grid on a rectangular domain.
269   !
270   !on exit
271   !===========
272   !X,Y        The positions on the uniform grid which are stored by
          indices of i,j
273   !
274   !Last change: Jan 17 2010
275   !-------------------------------------------------
276   SUBROUTINE  STRUCT_GRID (XI,ETA)
277   !-------------------------------------------------
278   USE PARAM,ONLY            : IM,JM,DXI,DETA
279   IMPLICIT NONE
280   INTEGER                   :: I,J
281   REAL(8),DIMENSION(IM,JM)  :: XI,ETA
282   !-------------------------------------------------
283   OPEN (UNIT=1,FILE='STRUCT_GRID.PLT',STATUS='UNKNOWN')
284   !-------------------------------------------------
285   DO J=1,JM
286      DO I=1,IM-1
287         XI(I+1,J)=XI(I,J)+DXI
288      END DO
289   END DO
290   DO I=1,IM
291      DO J=1,JM-1
292         ETA(I,J+1)=ETA(I,J)+DETA
293      END DO
294   END DO
295   !-------------------------------------------------
296   WRITE (1,*) 'VARIABLES=XI,ETA'
```

```
297 WRITE (1,*) 'ZONE'
298 WRITE (1,*) 'F=POINT'
299 WRITE (1,*) 'I=',IM
300 WRITE (1,*) 'J=',JM
301 DO I=1,IM
302    DO J=1,JM
303       WRITE (1,*) XI(I,J),ETA(I,J)
304    END DO
305 END DO
306 !--------------------------------------------------
307 WRITE(*,*) '******* Structured grid was written *******'
308 !--------------------------------------------------
309 END SUBROUTINE STRUCT_GRID
310 !*************************************************
311 !                 Neighbors
312 !*************************************************
313 SUBROUTINE NEIGHBORS (NODE_TYPE,L_NEIGHBOR,R_NEIGHBOR,U_NEIGHBOR,
        D_NEIGHBOR)
314 !--------------------------------------------------
315 USE PARAM,ONLY               : IM,NODE_NUM
316 IMPLICIT NONE
317 INTEGER                    :: INODE
318 LOGICAL,DIMENSION(NODE_NUM) :: NODE_TYPE
319 INTEGER,DIMENSION(NODE_NUM) :: L_NEIGHBOR,R_NEIGHBOR,U_NEIGHBOR,
        D_NEIGHBOR
320 !--------------------------------------------------
321 DO INODE=1,NODE_NUM
322    IF (.NOT.NODE_TYPE(INODE)) THEN
323       L_NEIGHBOR(INODE)=INODE-1
324       R_NEIGHBOR(INODE)=INODE+1
325       U_NEIGHBOR(INODE)=INODE+IM
326       D_NEIGHBOR(INODE)=INODE-IM
327    END IF
328 END DO
329 !--------------------------------------------------
330 WRITE(*,*) '******* Neighbors were found *******'
331 !--------------------------------------------------
332 END SUBROUTINE NEIGHBORS
333 !*************************************************
334 !        Boundary and Initial Conditions
335 !*************************************************
336 SUBROUTINE IBC (NODE_TYPE,T,OMEGA,F)
337 !--------------------------------------------------
338 USE PARAM   , ONLY                    : NODE_NUM
339 IMPLICIT NONE
340 INTEGER :: INODE
341 REAL(8),DIMENSION(NODE_NUM)           :: T,F,OMEGA
342 LOGICAL,DIMENSION(NODE_TYPE)          :: NODE_TYPE
343 !--------------------------------------------------
344 DO INODE=1,NODE_NUM
345    IF (    NODE_TYPE(INODE)) T(INODE)=1.D0
346    IF (.NOT.NODE_TYPE(INODE)) F(INODE)=-OMEGA(INODE)
347 END DO
348 !--------------------------------------------------
349 END SUBROUTINE IBC
350 !*************************************************
351 !           Krylov Subspace Solver
352 !*************************************************
353 SUBROUTINE SOLVERS (MD,NNZERO,RHS,RHS_POINTER,AA_AUX,JA_AUX,IA_AUX,T)
354 !--------------------------------------------------
355 USE PARAM , ONLY                    : IM,JM,NODE_NUM
356 IMPLICIT NONE
357 INTEGER                    :: NNZERO,I,MD
358 REAL(8),DIMENSION      (NODE_NUM)    :: T
359 REAL(8),DIMENSION      ((IM-2)*(JM-2)) :: RHS
360 INTEGER,DIMENSION      ((IM-2)*(JM-2)) :: RHS_POINTER
361 REAL(8),DIMENSION(MD)                 :: X0,R0,SOL
```

```fortran
362 REAL(8),DIMENSION(NNZERO)                    :: AA_AUX
363 INTEGER,DIMENSION(NNZERO)                    :: JA_AUX
364 INTEGER,DIMENSION(MD+1)                      :: IA_AUX
365 !------------------------------------------------
366 CALL INITIAL_GUESS (MD,X0)
367 !------------------------------------------------
368 CALL INITIAL_RESIDUAL (MD,NNZERO,AA_AUX,JA_AUX,IA_AUX,RHS,X0,R0)
369 !------------------------------------------------
370 ! Krylov subspaces methods
371 !------------------------------------------------
372 CALL CG                  (MD,NNZERO,AA_AUX,JA_AUX,IA_AUX,X0,R0,SOL)
373 !CALL ILU0_PCG            (MD,NNZERO,AA_AUX,JA_AUX,IA_AUX,X0,R0,SOL)
374 !CALL CGS                 (MD,NNZERO,AA_AUX,JA_AUX,IA_AUX,X0,R0,SOL)
375 !CALL ILU0_PCGS           (MD,NNZERO,AA_AUX,JA_AUX,IA_AUX,X0,R0,SOL)
376 !CALL BCG                 (MD,NNZERO,AA_AUX,JA_AUX,IA_AUX,X0,R0,SOL)
377 !CALL BCGSTAB             (MD,NNZERO,AA_AUX,JA_AUX,IA_AUX,X0,R0,SOL)
378 !CALL ILU0_PBCGSTAB       (MD,NNZERO,AA_AUX,JA_AUX,IA_AUX,X0,R0,SOL)
379 !CALL CR                  (MD,NNZERO,AA_AUX,JA_AUX,IA_AUX,X0,R0,SOL)
380 !CALL GMRES_STAR          (MD,NNZERO,AA_AUX,JA_AUX,IA_AUX,X0,R0,SOL)
381 !------------------------------------------------
382 DO I=1,MD
383    T(RHS_POINTER(I))=SOL(I)
384 END DO
385 !------------------------------------------------
386 END SUBROUTINE SOLVERS
387 !************************************************
388 !                RHS Constitution
389 !************************************************
390 SUBROUTINE RHS_CONSTITUTION (F,COEFF_C,COEFF_B,COEFF_L,COEFF_R,&
391 COEFF_T,L_POINT,R_POINT,T_POINT,B_POINT,NODE_TYPE,T,RHS_POINTER,RHS,MD,&
    NNZERO)
392 !------------------------------------------------
393 USE PARAM, ONLY          : IM,JM,NODE_NUM,DXI,DETA
394 IMPLICIT NONE
395 INTEGER                          :: INODE,K_RHS,NNZERO,MD
396 INTEGER,DIMENSION(NODE_NUM)      :: L_POINT,R_POINT,B_POINT,T_POINT
397 REAL(8),DIMENSION(NODE_NUM)      :: COEFF_C,COEFF_B,COEFF_L,COEFF_R
        ,COEFF_T
398 LOGICAL,DIMENSION(NODE_NUM)      :: NODE_TYPE
399 REAL(8),DIMENSION(NODE_NUM)      :: T,F
400 REAL(8),DIMENSION((IM-2)*(JM-2)) :: RHS
401 INTEGER,DIMENSION((IM-2)*(JM-2)) :: RHS_POINTER
402 !------------------------------------------------
403 K_RHS=0
404 NNZERO=0
405 DO INODE=1,NODE_NUM
406 !------------------------------------------------
407 ! Determines whether a node is boundary or not
408 !------------------------------------------------
409    IF (.NOT.NODE_TYPE(INODE)) THEN
410 !------------------------------------------------
411       K_RHS=K_RHS+1
412 !------------------------------------------------
413 ! Determines the RHS before any change
414 !------------------------------------------------
415       RHS(K_RHS)=F(INODE)
416 !------------------------------------------------
417 ! Determines the variable coefficient at the center node
418 !------------------------------------------------
419       COEFF_C(INODE)=-(2.D0/(DETA**2)+2.D0/(DXI**2))
420 !------------------------------------------------
421 ! Determines the variable coefficient at the bottom node
422 !------------------------------------------------
423       COEFF_B(B_POINT(INODE))=1.D0/(DETA**2)
424 !------------------------------------------------
425 ! Constitutes the new RHS on the basis whether bottom node is
426 ! boundary or not
```

```
427  !--------------------------------------------------------
428      IF (NODE_TYPE(B_POINT(INODE))) THEN
429      RHS(K_RHS)=RHS(K_RHS)-COEFF_B(B_POINT(INODE))*T(B_POINT(INODE))
430      ELSE
431      NNZERO=NNZERO+1
432      END IF
433  !--------------------------------------------------------
434  ! Determines the variable coefficient at the left node
435  !--------------------------------------------------------
436          COEFF_L(L_POINT(INODE))=1.D0/(DXI**2)
437  !--------------------------------------------------------
438  ! Constitutes the new RHS on the basis whether left node is
439  ! boundary or not
440  !--------------------------------------------------------
441          IF (NODE_TYPE(L_POINT(INODE))) THEN
442          RHS(K_RHS)=RHS(K_RHS)-COEFF_L(L_POINT(INODE))*T(L_POINT(INODE))
443          ELSE
444          NNZERO=NNZERO+1
445          END IF
446  !--------------------------------------------------------
447  ! Determines the variable coefficient at the right node
448  !--------------------------------------------------------
449          COEFF_R(R_POINT(INODE))=1.D0/(DXI**2)
450  !--------------------------------------------------------
451  ! Constitutes the new RHS on the basis whether right node is
452  ! boundary or not
453  !--------------------------------------------------------
454          IF (NODE_TYPE(R_POINT(INODE))) THEN
455          RHS(K_RHS)=RHS(K_RHS)-COEFF_R(R_POINT(INODE))*T(R_POINT(INODE))
456          ELSE
457          NNZERO=NNZERO+1
458          END IF
459  !--------------------------------------------------------
460  ! Determines the variable coefficient at the top node
461  !--------------------------------------------------------
462          COEFF_T(T_POINT(INODE))=1.D0/(DETA**2)
463  !--------------------------------------------------------
464  ! Constitutes the new RHS on the basis whether top node is
465  ! boundary or not
466  !--------------------------------------------------------
467          IF (NODE_TYPE(T_POINT(INODE))) THEN
468          RHS(K_RHS)=RHS(K_RHS)-COEFF_T(T_POINT(INODE))*T(T_POINT(INODE))
469          ELSE
470          NNZERO=NNZERO+1
471          END IF
472          RHS_POINTER(K_RHS)=INODE
473  !--------------------------------------------------------
474      END IF
475  END DO
476  NNZERO=NNZERO+K_RHS
477  MD=K_RHS
478  !--------------------------------------------------------
479  !WRITE(*,*) '******* RHS was constituted *******'
480  !--------------------------------------------------------
481  END SUBROUTINE RHS_CONSTITUTION
482  !************************************************
483  !              Matrix Constitution
484  !************************************************
485  SUBROUTINE MATRIX_CONSTITUTION (NNZERO,MD,COEFF_C,COEFF_B,COEFF_L,
         COEFF_R&
486  ,COEFF_T,NODE_TYPE,B_POINT,L_POINT,R_POINT,T_POINT,AA_AUX,
         JA_AUX_POINTER,IA_AUX,JA_AUX)
487  !--------------------------------------------------------
488  USE PARAM,ONLY           : NODE_NUM
489  IMPLICIT NONE
490  INTEGER                  :: INODE,CIR,IAA,ROW,MD,NNZERO,K,MBC,II
491  INTEGER,DIMENSION(MD+1)  :: IA_AUX
```

```
492 INTEGER,DIMENSION(NNZERO)          :: JA_AUX_POINTER,JA_AUX
493 REAL(8),DIMENSION(NNZERO)          :: AA_AUX
494 LOGICAL,DIMENSION(NODE_NUM)        :: NODE_TYPE
495 INTEGER,DIMENSION(NODE_NUM)        :: B_POINT,L_POINT,R_POINT,T_POINT
496 REAL(8),DIMENSION(NODE_NUM)        :: COEFF_C,COEFF_B,COEFF_L,COEFF_R,
        COEFF_T
497 !----------------------------------------------------
498 WRITE(*,*) 'Please wait, matrix constituting ...'
499 !----------------------------------------------------
500 IAA=0
501 CIR=0
502 ROW=1
503 IA_AUX(1)=1
504 DO INODE=1,NODE_NUM
505 IF (.NOT.NODE_TYPE(INODE)) THEN
506    ROW=ROW+1
507 !----------------------------------------------------
508    IF (.NOT.NODE_TYPE(B_POINT(INODE))) THEN
509       IAA=IAA+1
510       AA_AUX(IAA)=COEFF_B(B_POINT(INODE))
511       JA_AUX_POINTER(IAA)=B_POINT(INODE)
512       CIR=CIR+1
513    END IF
514 !----------------------------------------------------
515    IF (.NOT.NODE_TYPE(L_POINT(INODE))) THEN
516       IAA=IAA+1
517       AA_AUX(IAA)=COEFF_L(L_POINT(INODE))
518       JA_AUX_POINTER(IAA)=L_POINT(INODE)
519       CIR=CIR+1
520    END IF
521 !----------------------------------------------------
522    IAA=IAA+1
523    AA_AUX(IAA)=COEFF_C(INODE)
524    JA_AUX_POINTER(IAA)=INODE
525    CIR=CIR+1
526 !----------------------------------------------------
527    IF (.NOT.NODE_TYPE(R_POINT(INODE))) THEN
528       IAA=IAA+1
529       AA_AUX(IAA)=COEFF_R(R_POINT(INODE))
530       JA_AUX_POINTER(IAA)=R_POINT(INODE)
531       CIR=CIR+1
532    END IF
533 !----------------------------------------------------
534    IF (.NOT.NODE_TYPE(T_POINT(INODE))) THEN
535       IAA=IAA+1
536       AA_AUX(IAA)=COEFF_T(T_POINT(INODE))
537       JA_AUX_POINTER(IAA)=T_POINT(INODE)
538       CIR=CIR+1
539    END IF
540 !----------------------------------------------------
541 END IF
542 IA_AUX(ROW)=CIR+1
543 END DO
544 IA_AUX(MD+1)=IA_AUX(1)+NNZERO
545 !----------------------------------------------------
546 DO K=1,NNZERO
547 MBC=0
548 DO II=1,NODE_NUM
549 IF (NODE_TYPE(II).AND.II<JA_AUX_POINTER(K)) THEN
550 MBC=MBC+1
551 END IF
552 JA_AUX(K)=JA_AUX_POINTER(K)-MBC
553 END DO
554 END DO
555 !----------------------------------------------------
556 WRITE(*,*) '******* CSR format of matrix was constituted *******'
557 !----------------------------------------------------
```

```
558 END SUBROUTINE MATRIX_CONSTITUTION
559 !*************************************************
560 !                 Initial Guess
561 !*************************************************
562 SUBROUTINE INITIAL_GUESS (MD,X0)
563 !-----------------------------------------------
564 IMPLICIT NONE
565 INTEGER                         :: MD
566 REAL(8),DIMENSION(MD)           :: X0
567 !-----------------------------------------------
568 X0=0.D0
569 !-----------------------------------------------
570 END SUBROUTINE INITIAL_GUESS
571 !*************************************************
572 !                 Initial Residual
573 !*************************************************
574 SUBROUTINE INITIAL_RESIDUAL (MD,NNZERO,AA_AUX,JA_AUX,IA_AUX,RHS,X0,R0)
575 !-----------------------------------------------
576 IMPLICIT NONE
577 INTEGER                         :: MD,NNZERO
578 REAL(8),DIMENSION(MD)           :: RHS,X0,R0,Y0
579 REAL(8),DIMENSION(NNZERO)       :: AA_AUX
580 INTEGER,DIMENSION(NNZERO)       :: JA_AUX
581 INTEGER,DIMENSION(MD+1)         :: IA_AUX
582 !-----------------------------------------------
583 CALL CSR_MAT_V_PRODUCT (MD,NNZERO,AA_AUX,JA_AUX,IA_AUX,X0,Y0)
584 !-----------------------------------------------
585 R0=RHS-Y0
586 !-----------------------------------------------
587 END SUBROUTINE INITIAL_RESIDUAL
588 !*************************************************
589 !          Bi-Conjugate Gradiend Method
590 !*************************************************
591 SUBROUTINE BCG (MD,NNZERO,AA_AUX,JA_AUX,IA_AUX,X_OLD,R_OLD,X)
592 USE PARAM, ONLY              : TOL
593 IMPLICIT NONE
594 INTEGER                         :: K,ITER,MD,NNZERO
595 REAL(8),DIMENSION(MD)           :: X_OLD,X,P_OLD,P,R_OLD,R&
596 ,AP,RS_OLD,PS_OLD,PS,RS,ATPS
597 REAL(8),DIMENSION(NNZERO)       :: AA_AUX
598 INTEGER,DIMENSION(NNZERO)       :: JA_AUX
599 INTEGER,DIMENSION(MD+1)         :: IA_AUX
600 REAL(8)                         :: NORM,S,ALPHA,BETA,M,MM,SN
601 !-----------------------------------------------
602 RS_OLD  = R_OLD
603 P_OLD   = R_OLD
604 PS_OLD  = RS_OLD
605 NORM=1.D0
606 ITER=0
607 DO WHILE (NORM.GT.TOL)
608    ITER=ITER+1
609    !PRINT*,ITER
610    SN=0.D0
611    S=0.D0
612    M=0.D0
613    MM=0.D0
614 !-----------------------------------------------
615    CALL CSR_MAT_V_PRODUCT (MD,NNZERO,AA_AUX,JA_AUX,IA_AUX,P_OLD,AP)
616 !-----------------------------------------------
617    DO K=1,MD
618       M=M+R_OLD(K)*RS_OLD(K)
619       MM=MM+AP(K)*PS_OLD(K)
620    END DO
621    ALPHA=M/MM
622    X=X_OLD+ALPHA*P_OLD
623    R=R_OLD-ALPHA*AP
624 !-----------------------------------------------
```

```
625    CALL CSR_TMAT_V_PRODUCT (MD,NNZERO,AA_AUX,JA_AUX,IA_AUX,PS_OLD,ATPS)
626  !---------------------------------------------------
627    RS=RS_OLD-ALPHA*ATPS
628    DO K=1,MD
629       S=S+R(K)*RS(K)
630       SN=SN+R(K)*R(K)
631    END DO
632    NORM=SQRT(SN)/MD
633  !---------------------------------------------------
634    BETA=S/M
635    P=R+BETA*P_OLD
636    PS=RS+BETA*PS_OLD
637    PS_OLD   =PS
638    P_OLD    =P
639    X_OLD    =X
640    R_OLD    =R
641    RS_OLD   =RS
642    !PRINT*,NORM
643    !CALL SUCCESSIVE_SOL (NORM,ITER)
644  END DO
645  !---------------------------------------------------
646  END SUBROUTINE BCG
647  !**************************************************
648  !     CSR Format Matrix-Vector Multiplication
649  !**************************************************
650  SUBROUTINE CSR_MAT_V_PRODUCT (MD,NNZERO,AA_AUX,JA_AUX,IA_AUX,VE,YY)
651  !---------------------------------------------------
652  IMPLICIT NONE
653  INTEGER                         :: I,M1,M2,NNZERO,MD
654  REAL(8),DIMENSION(MD)           :: VE,YY
655  REAL(8),DIMENSION(NNZERO)       :: AA_AUX
656  INTEGER,DIMENSION(MD+1)         :: IA_AUX
657  INTEGER,DIMENSION(NNZERO)       :: JA_AUX
658  !---------------------------------------------------
659  YY=0.D0
660  DO I=1,MD
661     M1=IA_AUX(I)
662     M2=IA_AUX(I+1)-1
663     YY(I)=DOT_PRODUCT(AA_AUX(M1:M2),VE(JA_AUX(M1:M2)))
664  END DO
665  !---------------------------------------------------
666  END SUBROUTINE CSR_MAT_V_PRODUCT
667  !**************************************************
668  !CSR Format Transpose Matrix-Vector Multiplication
669  !**************************************************
670  SUBROUTINE CSR_TMAT_V_PRODUCT (MD,NNZERO,AA_AUX,JA_AUX,IA_AUX,VE,YY)
671  !---------------------------------------------------
672  IMPLICIT NONE
673  INTEGER                         :: I,K,NNZERO,MD
674  REAL(8),DIMENSION(MD)           :: VE,YY
675  REAL(8),DIMENSION(NNZERO)       :: AA_AUX
676  INTEGER,DIMENSION(MD+1)         :: IA_AUX
677  INTEGER,DIMENSION(NNZERO)       :: JA_AUX
678  !---------------------------------------------------
679  DO I=1,MD
680     YY(I)=0.D0
681  END DO
682  !---------------------------------------------------
683  ! loop over the rows
684  !---------------------------------------------------
685  DO I=1,MD
686     DO K=IA_AUX(I),IA_AUX(I+1)-1
687        YY(JA_AUX(K))=YY(JA_AUX(K))+VE(I)*AA_AUX(K)
688     END DO
689  END DO
690  !---------------------------------------------------
691  END SUBROUTINE CSR_TMAT_V_PRODUCT
```

```
692  !****************************************************
693  !     MSR Format Matrix-Vector Multiplication
694  !****************************************************
695  SUBROUTINE MSR_MAT_V_PRODUCT (TL,MD,SA,IJA,X,B)
696  !----------------------------------------------------
697  IMPLICIT NONE
698  INTEGER                          :: I,K,TL,MD
699  REAL(8),DIMENSION(MD)            :: B,X
700  REAL(8),DIMENSION(TL)            :: SA
701  INTEGER,DIMENSION(TL)            :: IJA
702  !----------------------------------------------------
703  IF (IJA(1).NE.MD+2) PAUSE 'mismatched vector and matrix in
         MSR_MAT_V_PRODUCT'
704  DO I=1,MD
705     B(I)=SA(I)*X(I)
706     DO K=IJA(I),IJA(I+1)-1
707        B(I)=B(I)+SA(K)*X(IJA(K))
708     END DO
709  END DO
710  !----------------------------------------------------
711  END SUBROUTINE MSR_MAT_V_PRODUCT
712  !****************************************************
713  !MSR Format Transpose Matrix-Vector Multiplication
714  !****************************************************
715  SUBROUTINE MSR_TMAT_V_PRODUCT (TL,MD,SA,IJA,X,B)
716  IMPLICIT NONE
717  INTEGER                          :: I,J,K,TL,MD
718  REAL(8),DIMENSION(MD)            :: B,X
719  INTEGER,DIMENSION(TL)            :: IJA
720  REAL(8),DIMENSION(TL)            :: SA
721  !----------------------------------------------------
722  IF (IJA(1).NE.MD+2) PAUSE 'mismatched vector and matrix in
         MSR_TMAT_V_PRODUCT'
723  DO I=1,MD
724     B(I)=SA(I)*X(I)
725  END DO
726  DO I=1,MD
727     DO K=IJA(I),IJA(I+1)-1
728        J=IJA(K)
729        B(J)=B(J)+SA(K)*X(I)
730     END DO
731  END DO
732  !----------------------------------------------------
733  END SUBROUTINE MSR_TMAT_V_PRODUCT
734  !****************************************************
735  !        Total Length of MSR Format Vectors
736  !****************************************************
737  SUBROUTINE TOTAL_LENGTH (MD,IA,TL)
738  !----------------------------------------------------
739  IMPLICIT NONE
740  INTEGER                          :: TL,MD
741  INTEGER,DIMENSION(MD+1)          :: IA
742  !----------------------------------------------------
743  TL=IA(MD+1)
744  !----------------------------------------------------
745  END SUBROUTINE TOTAL_LENGTH
746  !****************************************************
747  !              CSR to MSR Conversion
748  !****************************************************
749  SUBROUTINE CSRMSR   (TL,NROW,NNZ,AA,JA,IA,AO,JAO)
750  !----------------------------------------------------
751  IMPLICIT NONE
752  INTEGER                          :: I,J,K,NNZ,II,TL,NROW
753  REAL(8),DIMENSION(NNZ)           :: AA
754  REAL(8),DIMENSION(TL)            :: AO
755  REAL(8),DIMENSION(NROW)          :: WK
756  INTEGER,DIMENSION(NROW+1)        :: IA,IWK
```

```
757 INTEGER,DIMENSION(NNZ)                :: JA
758 INTEGER,DIMENSION(TL)                 :: JAO
759 !-------------------------------------------------
760 DO  I=1,NROW
761     WK(I) = 0.D0
762     IWK(I+1) = IA(I+1)-IA(I)
763     DO  K=IA(I),IA(I+1)-1
764         IF (JA(K).EQ.I) THEN
765             WK(I)= AA(K)
766             IWK(I+1)=IWK(I+1)-1
767         END IF
768     END DO
769 END DO
770 !-------------------------------------------------
771 ! Copy backwards (to avoid collisions)
772 !-------------------------------------------------
773 DO II=NROW,1,-1
774     DO K=IA(II+1)-1,IA(II),-1
775         J=JA(K)
776         IF (J.NE.II) THEN
777             AO(TL) = AA(K)
778             JAO(TL) = J
779             TL = TL-1
780         END IF
781     END DO
782 END DO
783 !-------------------------------------------------
784 ! Compute pointer values and copy WK(NROW)
785 !-------------------------------------------------
786 JAO(1)=NROW+2
787 DO  I=1,NROW
788     AO(I)=WK(I)
789     JAO(I+1)=JAO(I)+IWK(I+1)
790 END DO
791 !-------------------------------------------------
792 END SUBROUTINE CSRMSR
793 !**************************************************
794 !              Successive Solution
795 !**************************************************
796 SUBROUTINE SUCCESSIVE_SOL (ERR,ITER)
797 !-------------------------------------------------
798 IMPLICIT NONE
799 INTEGER                           :: ITER,COUNTER
800 CHARACTER*10                      :: EXT
801 CHARACTER*4                       :: FN1
802 CHARACTER*25                      :: FNAME
803 REAL(8)                           :: ERR
804 !-------------------------------------------------
805 FN1='VECT'
806 COUNTER=0
807 !-------------------------------------------------
808 COUNTER=COUNTER+10
809 WRITE(EXT,'(I7)') ITER
810 FNAME=FN1//EXT//'.DAT'
811 !-------------------------------------------------
812 OPEN(COUNTER,FILE=FNAME,POSITION='REWIND')
813 !WRITE(COUNTER,*)'VARIABLES= "ITER","ERR"'
814 WRITE(COUNTER,*) ITER,ERR
815 CLOSE(COUNTER)
816 !-------------------------------------------------
817 !WRITE(*,*) '========================'
818 !WRITE(*,*) 'PRINTING ON ',FNAME
819 !WRITE(*,*) '========================'
820 !-------------------------------------------------
821 END SUBROUTINE SUCCESSIVE_SOL
822 !**************************************************
823 !              Conjugate Gradien Method
```

```
824  !*****************************************************
825  SUBROUTINE CG (MD,NNZERO,AA_AUX,JA_AUX,IA_AUX,X_OLD,R_OLD,X)
826  USE PARAM, ONLY                       : TOL
827  IMPLICIT NONE
828  INTEGER                               :: K,ITER,TL,MD,NNZERO,IDIAG,IERR
829  REAL(8),DIMENSION(MD)                 :: X_OLD,X,P_OLD,P,R_OLD,R,AP
830  INTEGER,DIMENSION(MD+1)               :: IA_AUX,IAR
831  INTEGER,DIMENSION(2*MD-1)             :: IND
832  REAL(8)                               :: NORM,S,ALPHA,BETA,M,MM
833  REAL(8),DIMENSION(NNZERO)             :: AA_AUX,AAR
834  INTEGER,DIMENSION(NNZERO)             :: JA_AUX,JAR
835  REAL(8),ALLOCATABLE                   :: SA(:),DIAG(:,:),COEF(:,:)
836  INTEGER,ALLOCATABLE                   :: IJA(:),IOFF(:),JCOEF(:,:)
837  !-------------------------------------------------
838  CALL TOTAL_LENGTH        (MD,IA_AUX,TL)
839  ALLOCATE                 (SA(TL),IJA(TL))
840  CALL CSRMSR              (TL,MD,NNZERO,AA_AUX,JA_AUX,IA_AUX,SA,IJA)
841  CALL INFDIA              (MD,NNZERO,JA_AUX,IA_AUX,IND,IDIAG)
842  ALLOCATE                 (COEF(MD,IDIAG),JCOEF(MD,IDIAG))
843  ALLOCATE                 (IOFF(IDIAG),DIAG(MD,IDIAG))
844  CALL CSRELL              (MD,NNZERO,AA_AUX,JA_AUX,IA_AUX,MD,COEF,
         JCOEF,MD,IDIAG,IERR)
845  CALL CSRDIA             (MD,NNZERO,IDIAG,10,AA_AUX,JA_AUX,IA_AUX,MD,
         DIAG,IOFF,AAR,JAR,IAR,IND)
846  !-------------------------------------------------
847  P_OLD=R_OLD
848  NORM=1.D0
849  ITER=0
850  DO WHILE (NORM.GT.TOL)
851     ITER=ITER+1
852     !PRINT*,ITER
853     S=0.D0
854     M=0.D0
855     MM=0.D0
856  !-------------------------------------------------
857  ! Stores the matrix in compressed spars row and multiplies by
858  ! vector
859  !-------------------------------------------------
860     !CALL CSR_MAT_V_PRODUCT    (MD,NNZERO,AA_AUX,JA_AUX,IA_AUX,P_OLD,AP
         )
861  !-------------------------------------------------
862  ! Stores the matrix in modified spars row and multiplies by
863  ! vector
864  !-------------------------------------------------
865     !CALL MSR_MAT_V_PRODUCT (TL,MD,SA,IJA,P_OLD,AP)
866  !-------------------------------------------------
867  ! Stores the matrix in Ellpack/Itpack format and multiplies by
868  ! vector
869  !-------------------------------------------------
870     !CALL ELLPACK_ITPACK_MAT_V_PRODUCT (MD,P_OLD,AP,MD,IDIAG,COEF,JCOEF)
871  !-------------------------------------------------
872  ! Stores the matrix in Diagonal format and multiplies by
873  ! vector
874  !-------------------------------------------------
875     CALL DIAGONAL_MAT_V_PRODUCT (MD,P_OLD,AP,DIAG,MD,IDIAG,IOFF)
876  !-------------------------------------------------
877     DO K=1,MD
878        M=M+R_OLD(K)**2
879        MM=MM+AP(K)*P_OLD(K)
880     END DO
881     ALPHA=M/MM
882     X=X_OLD+ALPHA*P_OLD
883     R=R_OLD-ALPHA*AP
884  !-------------------------------------------------
885  ! Computing the Euclidean norm
886  !-------------------------------------------------
887     DO K=1,MD
```

```
888          S=S+R(K)**2
889     END DO
890     NORM=SQRT(S)/MD
891  !-------------------------------------------------
892     BETA=S/M
893     P=R+BETA*P_OLD
894     P_OLD=P
895     X_OLD=X
896     R_OLD=R
897     !PRINT*,NORM
898     !CALL SUCCESSIVE_SOL (NORM,ITER)
899  END DO
900  !-------------------------------------------------
901  END SUBROUTINE CG
902  !**********************************************************
903  !   Incomplete LU Factorization with 0 Level of Fill in
904  !**********************************************************
905  SUBROUTINE ILU0 (N,NNZERO,A,JA,IA,LUVAL,UPTR,IW,ICODE)
906  !-------------------------------------------------
907  IMPLICIT NONE
908  INTEGER                      :: N,I,J,K,J1,J2,ICODE,JROW,NNZERO,JJ,JW
909  INTEGER,DIMENSION(NNZERO)    :: JA
910  INTEGER,DIMENSION(N+1)       :: IA
911  INTEGER,DIMENSION(N)         :: IW,UPTR
912  REAL(8),DIMENSION(NNZERO)    :: A
913  REAL(8),DIMENSION(NNZERO)    :: LUVAL
914  REAL(8)                      :: T
915  !-------------------------------------------------
916  DO I=1,IA(N+1)-1
917     LUVAL(I)=A(I)
918  END DO
919  DO I=1,N
920     IW(I)=0
921  END DO
922  !-------------------------------------------------
923  DO K=1,N
924     J1=IA(K)
925     J2=IA(K+1)-1
926     DO J=J1,J2
927        IW(JA(J))=J
928     END DO
929        J=J1
930  150   JROW=JA(J)
931  !-------------------------------------------------
932     IF (JROW.GE.K) GO TO 200
933  !-------------------------------------------------
934     T=LUVAL(J)*LUVAL(UPTR(JROW))
935     LUVAL(J)=T
936  !-------------------------------------------------
937     DO JJ=UPTR(JROW)+1,IA(JROW+1)-1
938        JW=IW(JA(JJ))
939        IF (JW.NE.0) LUVAL(JW)=LUVAL(JW)-T*LUVAL(JJ)
940     END DO
941     J=J+1
942     IF (J.LE.J2) GO TO 150
943  !-------------------------------------------------
944  200   UPTR(K)=J
945     IF (JROW.NE.K.OR.LUVAL(J).EQ.0.D0) GO TO 600
946     LUVAL(J)=1.D0/LUVAL(J)
947  !-------------------------------------------------
948     DO I=J1,J2
949        IW(JA(I))=0
950     END DO
951  END DO
952  !-------------------------------------------------
953  ICODE=0
954  RETURN
```

```
955 !-----------------------------------------------
956 600   ICODE=K
957 RETURN
958 !-----------------------------------------------
959 END SUBROUTINE ILU0
960 !************************************************************
961 !                      LU Solution
962 !************************************************************
963 SUBROUTINE LUSOL  (N,NNZERO,RHS,SOL,LUVAL,LUCOL,LUPTR,UPTR)
964 !-----------------------------------------------
965 IMPLICIT NONE
966 INTEGER                       :: I,K,N,NNZERO
967 REAL(8),DIMENSION(NNZERO)     :: LUVAL
968 REAL(8),DIMENSION(N)          :: SOL,RHS
969 INTEGER,DIMENSION(N)          :: UPTR
970 INTEGER,DIMENSION(N+1)        :: LUPTR
971 INTEGER,DIMENSION(NNZERO)     :: LUCOL
972 !-----------------------------------------------
973 DO I=1,N
974    SOL(I)=RHS(I)
975    DO K=LUPTR(I),UPTR(I)-1
976       SOL(I)=SOL(I)-LUVAL(K)*SOL(LUCOL(K))
977    END DO
978 END DO
979 !-----------------------------------------------
980 DO I=N,1,-1
981    DO K=UPTR(I)+1,LUPTR(I+1)-1
982       SOL(I)=SOL(I)-LUVAL(K)*SOL(LUCOL(K))
983    END DO
984    SOL(I)=LUVAL(UPTR(I))*SOL(I)
985 END DO
986 !-----------------------------------------------
987 END SUBROUTINE LUSOL
988 !************************************************************
989 !ILU(0)-Preconditioned Conjugate Gradient Method
990 !************************************************************
991 SUBROUTINE ILU0_PCG (N,NNZERO,AA_AUX,JA_AUX,IA_AUX,X_OLD,R_OLD,X)
992 !-----------------------------------------------
993 USE PARAM, ONLY              : TOL
994 IMPLICIT NONE
995 INTEGER                     :: K,ITER,TL,N,NNZERO,ICODE,IDIAG,IERR
996 REAL(8),DIMENSION(N)        :: X_OLD,X,P_OLD,P,R_OLD,R,AP,Z_OLD,Z
997 INTEGER,DIMENSION(N+1)      :: IA_AUX,IAR
998 INTEGER,DIMENSION(2*N-1)    :: IND
999 REAL(8)                     :: NORM,S,ALPHA,BETA,M,MM
1000 REAL(8),DIMENSION(NNZERO)  :: AA_AUX,AAR,LUVAL
1001 INTEGER,DIMENSION(NNZERO)  :: JA_AUX,JAR
1002 REAL(8),ALLOCATABLE        :: SA(:),COEF(:,:),DIAG(:,:)
1003 INTEGER,ALLOCATABLE        :: IJA(:),JCOEF(:,:),IOFF(:)
1004 INTEGER,DIMENSION(N)       :: UPTR,IW
1005 REAL(8)                    :: SN
1006 !-----------------------------------------------
1007 CALL TOTAL_LENGTH  (N,IA_AUX,TL)
1008 ALLOCATE          (SA(TL),IJA(TL))
1009 CALL CSRMSR        (TL,N,NNZERO,AA_AUX,JA_AUX,IA_AUX,SA,IJA)
1010 CALL INFDIA        (N,NNZERO,JA_AUX,IA_AUX,IND,IDIAG)
1011 ALLOCATE          (COEF(N,IDIAG),JCOEF(N,IDIAG))
1012 ALLOCATE          (IOFF(IDIAG),DIAG(N,IDIAG))
1013 CALL CSRELL        (N,NNZERO,AA_AUX,JA_AUX,IA_AUX,N,COEF,JCOEF,N,IDIAG,
                         IERR)
1014 CALL CSRDIA        (N,NNZERO,IDIAG,10,AA_AUX,JA_AUX,IA_AUX,N,DIAG,IOFF,
                         AAR,JAR,IAR,IND)
1015 !-----------------------------------------------
1016 CALL ILU0          (N,NNZERO,AA_AUX,JA_AUX,IA_AUX,LUVAL,UPTR,IW,ICODE)
1017 CALL LUSOL         (N,NNZERO,R_OLD,Z_OLD,LUVAL,JA_AUX,IA_AUX,UPTR)
1018 !-----------------------------------------------
1019 P_OLD=Z_OLD
```

```
1020  NORM=1.D0
1021  ITER=0
1022  DO WHILE (NORM.GT.TOL)
1023     ITER=ITER+1
1024     !PRINT*,ITER
1025     SN=0.D0
1026     S=0.D0
1027     M=0.D0
1028     MM=0.D0
1029  !-----------------------------------------------
1030  ! Stores the matrix in compressed spars row and multiplies by
1031  ! vector
1032  !-----------------------------------------------
1033  !CALL CSR_MAT_V_PRODUCT     (N,NNZERO,AA_AUX,JA_AUX,IA_AUX,P_OLD,AP)
1034  !-----------------------------------------------
1035  ! Stores the matrix in modified spars row and multiplies by
1036  ! vector
1037  !-----------------------------------------------
1038  !CALL MSR_MAT_V_PRODUCT (TL,N,SA,IJA,P_OLD,AP)
1039  !-----------------------------------------------
1040  ! Stores the matrix in Ellpack/Itpack format and multiplies by
1041  ! vector
1042  !-----------------------------------------------
1043  !CALL ELLPACK_ITPACK_MAT_V_PRODUCT (N,P_OLD,AP,N,IDIAG,COEF,JCOEF)
1044  !-----------------------------------------------
1045  ! Stores the matrix in Diagonal format and multiplies by
1046  ! vector
1047  !-----------------------------------------------
1048     CALL DIAGONAL_MAT_V_PRODUCT (N,P_OLD,AP,DIAG,N,IDIAG,IOFF)
1049  !-----------------------------------------------
1050     DO  K=1,N
1051        M=M+R_OLD(K)*Z_OLD(K)
1052        MM=MM+AP(K)*P_OLD(K)
1053     END DO
1054     ALPHA=M/MM
1055     X=X_OLD+ALPHA*P_OLD
1056     R=R_OLD-ALPHA*AP
1057  !-----------------------------------------------
1058     CALL LUSOL    (N,NNZERO,R,Z,LUVAL,JA_AUX,IA_AUX,UPTR)
1059  !-----------------------------------------------
1060     DO  K=1,N
1061        S=S+R(K)*Z(K)
1062     END DO
1063     BETA=S/M
1064     P=Z+BETA*P_OLD
1065     DO  K=1,N
1066        SN=SN+R(K)**2
1067     END DO
1068     NORM=SQRT(SN)/N
1069  !-----------------------------------------------
1070     P_OLD=P
1071     X_OLD=X
1072     R_OLD=R
1073     Z_OLD=Z
1074     !PRINT*,NORM
1075     !CALL SUCCESSIVE_SOL (NORM,ITER)
1076  END DO
1077  !-----------------------------------------------
1078  END SUBROUTINE ILU0_PCG
1079  !*************************************************
1080  !                 CPU Time Write
1081  !*************************************************
1082  SUBROUTINE  CPU_TIME_WRITE (TIME_BEGIN,TIME_END,ITER)
1083  !-----------------------------------------------
1084  IMPLICIT NONE
1085  INTEGER                              :: ITER,COUNTER
1086  CHARACTER*10                         :: EXT
```

```
1087 CHARACTER*5                              :: FN1
1088 CHARACTER*25                             :: FNAME
1089 REAL(8)                                  :: TIME_BEGIN,TIME_END
1090 !-----------------------------------------------------
1091 FN1='CPU_T'
1092 COUNTER=0
1093 !-----------------------------------------------------
1094 COUNTER=COUNTER+10
1095 WRITE(EXT,'(I7)') ITER
1096 FNAME=FN1//EXT//'.DAT'
1097 !-----------------------------------------------------
1098 OPEN(COUNTER,FILE=FNAME,POSITION='REWIND')
1099 !WRITE(COUNTER,*)'VARIABLES= "ITER","TIME"'
1100 !WRITE(COUNTER,*)'ZONE,F=POINT'
1101 WRITE(COUNTER,*) ITER,TIME_END-TIME_BEGIN
1102 CLOSE(COUNTER)
1103 !-----------------------------------------------------
1104 !WRITE(*,*) '========================='
1105 !WRITE(*,*) 'PRINTING ON ',FNAME
1106 !WRITE(*,*) '========================='
1107 !-----------------------------------------------------
1108 END SUBROUTINE CPU_TIME_WRITE
1109 !**************************************************
1110 !                  INFDIA
1111 !**************************************************
1112 SUBROUTINE INFDIA (MD,NNZ,JA,IA,IND,IDIAG)
1113 !-----------------------------------------------------
1114 IMPLICIT NONE
1115 INTEGER                       :: MD,NNZ,N2,I,K,J,IDIAG
1116 INTEGER,DIMENSION(MD+1)        :: IA
1117 INTEGER,DIMENSION(2*MD-1)      :: IND
1118 INTEGER,DIMENSION(NNZ)         :: JA
1119 !-----------------------------------------------------
1120 N2=MD+MD-1
1121 DO I=1,N2
1122    IND(I)=0
1123 END DO
1124 DO I=1,MD
1125    DO K=IA(I),IA(I+1)-1
1126       J=JA(K)
1127       IND(MD+J-I)=IND(MD+J-I)+1
1128    END DO
1129 END DO
1130 !-----------------------------------------------------
1131 !     count the nonzero ones.
1132 !-----------------------------------------------------
1133 IDIAG=0
1134 DO K=1,N2
1135    IF (IND(K).NE.0) IDIAG=IDIAG+1
1136 END DO
1137 !-----------------------------------------------------
1138 END SUBROUTINE INFDIA
1139 !**************************************************
1140 !      CSR to Ellpack-Itpack Conversion
1141 !**************************************************
1142 SUBROUTINE CSRELL (NROW,NN,A,JA,IA,MAXCOL,COEF,JCOEF,NCOEF,NDIAG,IERR)
1143 IMPLICIT NONE
1144 INTEGER                :: NROW,NN,NCOEF,NDIAG,IERR,I,J,K,K1,K2,MAXCOL
1145 INTEGER,DIMENSION(NROW+1)      :: IA
1146 INTEGER,DIMENSION(NN)          :: JA
1147 INTEGER,DIMENSION(NCOEF,NDIAG) :: JCOEF
1148 REAL(8),DIMENSION(NN)          :: A
1149 REAL(8),DIMENSION(NCOEF,NDIAG) :: COEF
1150 !-----------------------------------------------------
1151 ! first determine the length of each row of lower-part-of(A)
1152 !-----------------------------------------------------
1153 IERR=0
```

```
1154 NDIAG=0
1155 DO  I=1,NROW
1156     K=IA(I+1)-IA(I)
1157     NDIAG=MAX0(NDIAG,K)
1158 END DO
1159 !-------------------------------------------------
1160 ! check whether sufficient columns are available.
1161 !-------------------------------------------------
1162 IF (NDIAG.GT.MAXCOL) THEN
1163     IERR=1
1164 ENDIF
1165 !-------------------------------------------------
1166 ! fill COEF with zero elements and jcoef with row numbers.
1167 !-------------------------------------------------
1168 DO  J=1,NDIAG
1169     DO  I=1,NROW
1170         COEF(I,J)=0.D0
1171         JCOEF(I,J)=I
1172     END DO
1173 END DO
1174 !-------------------------------------------------
1175 !        copy elements row by row.
1176 !-------------------------------------------------
1177 DO  I=1,NROW
1178     K1=IA(I)
1179     K2=IA(I+1)-1
1180     DO  K=K1,K2
1181         COEF(I,K-K1+1)=A(K)
1182         JCOEF(I,K-K1+1)=JA(K)
1183     END DO
1184 END DO
1185 !-------------------------------------------------
1186 END SUBROUTINE CSRELL
1187 !*************************************************************
1188 ! Ellpack-Itpack Format Matrix-Vector Multiplication
1189 !*************************************************************
1190 SUBROUTINE ELLPACK_ITPACK_MAT_V_PRODUCT (N,X,Y,NA,NCOL,A,JA)
1191 !-------------------------------------------------
1192 IMPLICIT NONE
1193 INTEGER                              :: N,NA,NCOL,I,J
1194 REAL(8),DIMENSION(N)                 :: X,Y
1195 REAL(8),DIMENSION(NA,NCOL)           :: A
1196 INTEGER,DIMENSION(NA,NCOL)           :: JA
1197 !-------------------------------------------------
1198 DO I=1,N
1199     Y(I)=0.D0
1200 END DO
1201 DO J=1,NCOL
1202     DO  I=1,N
1203         Y(I)=Y(I)+A(I,J)*X(JA(I,J))
1204     END DO
1205 END DO
1206 !-------------------------------------------------
1207 END SUBROUTINE ELLPACK_ITPACK_MAT_V_PRODUCT
1208 !*************************************************************
1209 !       CSR to Diagonal Conversion
1210 !*************************************************************
1211 SUBROUTINE CSRDIA (MD,NNZ,IDIAG,JOB,A,JA,IA,NDIAG,DIAG,IOFF,AO,JAO,IAO,
          IND)
1212 !-------------------------------------------------
1213 IMPLICIT NONE
1214 INTEGER                              :: MD,NNZ,IDIAG,NDIAG,JOB1,JOB2,JOB,
          N2,IDUM,II,JMAX,K,I,J,L,KO
1215 REAL(8),DIMENSION(NDIAG,IDIAG)       :: DIAG
1216 REAL(8),DIMENSION(NNZ)               :: A,AO
1217 INTEGER,DIMENSION(NNZ)               :: JA,JAO
1218 INTEGER,DIMENSION(MD+1)              :: IA,IAO
```

```
1219 INTEGER,DIMENSION(2*MD-1)            :: IND
1220 INTEGER,DIMENSION(IDIAG)             :: IOFF
1221 !------------------------------------------------
1222 JOB1 = JOB/10
1223 JOB2 = JOB-JOB1*10
1224 IF (JOB1 .NE. 0) THEN
1225     N2 = MD+MD-1
1226     CALL INFDIA(MD,NNZ,JA,IA,IND,IDUM)
1227 !------------------------------------------------
1228 ! determine diagonals to  accept.
1229 !------------------------------------------------
1230     II=0
1231     DO
1232         II=II+1
1233         JMAX=0
1234         DO  K=1,N2
1235             J=IND(K)
1236             IF (JMAX.LT.J) THEN
1237                 I=K
1238                 JMAX=J
1239             END IF
1240         END DO
1241         IF (JMAX.LE.0) THEN
1242             II=II-1
1243             EXIT
1244         ENDIF
1245         IOFF(II)=I-MD
1246         IND(I)=-JMAX
1247         IF (IDIAG.LE.II) THEN
1248             EXIT
1249         END IF
1250     END DO
1251     IDIAG=II
1252 END IF
1253 !------------------------------------------------
1254 ! initialize diago to zero
1255 !------------------------------------------------
1256 DO  J=1,IDIAG
1257     DO  I=1,MD
1258         DIAG(I,J)=0.D0
1259     END DO
1260 END DO
1261 KO=1
1262 !------------------------------------------------
1263 ! extract diagonals and accumulate remaining matrix.
1264 !------------------------------------------------
1265 DO  I=1,MD
1266     DO  K=IA(I),IA(I+1)-1
1267         J=JA(K)
1268         DO L=1,IDIAG
1269             IF (J-I.EQ.IOFF(L)) THEN
1270                 DIAG(I,L)=A(K)
1271                 GOTO 51
1272             END IF
1273         END DO
1274 !------------------------------------------------
1275 ! append element not in any diagonal to AO,JAO,IAO
1276 !------------------------------------------------
1277         IF (JOB2.NE.0) THEN
1278             AO(KO)=A(K)
1279             JAO(KO)=J
1280             KO=KO+1
1281         END IF
1282 51      CONTINUE
1283     END DO
1284     IF (JOB2.NE.0) THEN
1285         IND(I+1)=KO
```

```fortran
1286          END IF
1287 END DO
1288 !---------------------------------------------------
1289 !      finish with IAO
1290 !---------------------------------------------------
1291 IF (JOB2.NE.0) THEN
1292     IAO(1)=1
1293     DO I=2,MD+1
1294         IAO(I)=IND(I)
1295     END DO
1296 END IF
1297 !---------------------------------------------------
1298 END SUBROUTINE CSRDIA
1299 !**************************************************
1300 ! Diagonal Format Matrix-Vector Multiplication
1301 !**************************************************
1302 SUBROUTINE DIAGONAL_MAT_V_PRODUCT (MD,X,Y,DIAG,NDIAG,IDIAG,IOFF)
1303 !---------------------------------------------------
1304 IMPLICIT NONE
1305 INTEGER                       :: MD,J,K,IO,I1,I2,NDIAG,IDIAG
1306 INTEGER,DIMENSION(IDIAG)      :: IOFF
1307 REAL(8),DIMENSION(MD)         :: X,Y
1308 REAL(8),DIMENSION(NDIAG,IDIAG) :: DIAG
1309 !---------------------------------------------------
1310 DO J=1,MD
1311     Y(J)=0.D0
1312 END DO
1313 DO J=1,IDIAG
1314     IO=IOFF(J)
1315     I1=MAX0(1,1-IO)
1316     I2=MIN0(MD,MD-IO)
1317     DO K=I1,I2
1318         Y(K)=Y(K)+DIAG(K,J)*X(K+IO)
1319     END DO
1320 END DO
1321 !---------------------------------------------------
1322 END SUBROUTINE DIAGONAL_MAT_V_PRODUCT
1323 !**************************************************
1324 !         Conjugate Gradien Squared Method
1325 !**************************************************
1326 SUBROUTINE CGS (N,NNZERO,AA_AUX,JA_AUX,IA_AUX,X_OLD,R_OLD,X)
1327 !---------------------------------------------------
1328 USE PARAM, ONLY             : TOL
1329 IMPLICIT NONE
1330 INTEGER                     :: K,ITER,N,NNZERO,TL,IDIAG,IERR
1331 REAL(8),DIMENSION(NNZERO)   :: AA_AUX,AAR
1332 INTEGER,DIMENSION(NNZERO)   :: JA_AUX,JAR
1333 INTEGER,DIMENSION(2*N-1)    :: IND
1334 INTEGER,DIMENSION(N+1)      :: IA_AUX,IAR
1335 REAL(8),ALLOCATABLE         :: SA(:),COEF(:,:),DIAG(:,:)
1336 INTEGER,ALLOCATABLE         :: IJA(:),JCOEF(:,:),IOFF(:)
1337 REAL(8),DIMENSION(N)        :: X_OLD,X,P_OLD,P,R_OLD,U,U_OLD,R,
        AP,AUQ,RS,Q
1338 REAL(8)                     :: NORM,S,ALPHA,BETA,M,MM,MN
1339 !---------------------------------------------------
1340 CALL TOTAL_LENGTH          (N,IA_AUX,TL)
1341 ALLOCATE                   (SA(TL),IJA(TL))
1342 CALL CSRMSR                (TL,N,NNZERO,AA_AUX,JA_AUX,IA_AUX,SA,IJA)
1343 CALL INFDIA                (N,NNZERO,JA_AUX,IA_AUX,IND,IDIAG)
1344 ALLOCATE                   (COEF(N,IDIAG),JCOEF(N,IDIAG))
1345 ALLOCATE                   (IOFF(IDIAG),DIAG(N,IDIAG))
1346 CALL CSRELL                (N,NNZERO,AA_AUX,JA_AUX,IA_AUX,N,COEF,JCOEF,
        N,IDIAG,IERR)
1347 CALL CSRDIA                (N,NNZERO,IDIAG,10,AA_AUX,JA_AUX,IA_AUX,N,
        DIAG,IOFF,AAR,JAR,IAR,IND)
1348 !---------------------------------------------------
1349 RS=R_OLD
```

```
1350  P_OLD=R_OLD
1351  U_OLD=R_OLD
1352  NORM=1.D0
1353  ITER=0
1354  DO WHILE (NORM.GT.TOL)
1355     ITER=ITER+1
1356     !PRINT*,ITER
1357     S=0.D0
1358     M=0.D0
1359     MM=0.D0
1360     MN=0.D0
1361  !-------------------------------------------------------
1362  ! Stores the matrix in compressed spars row and multiplies by
1363  ! vector
1364  !-------------------------------------------------------
1365     !CALL CSR_MAT_V_PRODUCT      (N,NNZERO,AA_AUX,JA_AUX,IA_AUX,P_OLD,AP)
1366  !-------------------------------------------------------
1367  ! Stores the matrix in modified spars row and multiplies by
1368  ! vector
1369  !-------------------------------------------------------
1370     !CALL MSR_MAT_V_PRODUCT (TL,N,SA,IJA,P_OLD,AP)
1371  !-------------------------------------------------------
1372  ! Stores the matrix in Ellpack/Itpack format and multiplies by
1373  ! vector
1374  !-------------------------------------------------------
1375     !CALL ELLPACK_ITPACK_MAT_V_PRODUCT (N,P_OLD,AP,N,IDIAG,COEF,JCOEF)
1376  !-------------------------------------------------------
1377  ! Stores the matrix in Diagonal format and multiplies by
1378  ! vector
1379  !-------------------------------------------------------
1380     CALL DIAGONAL_MAT_V_PRODUCT (N,P_OLD,AP,DIAG,N,IDIAG,IOFF)
1381  !-------------------------------------------------------
1382     DO K=1,N
1383        M=M+R_OLD(K)*RS(K)
1384        MM=MM+AP(K)*RS(K)
1385     END DO
1386     ALPHA=M/MM
1387     Q=U_OLD-ALPHA*AP
1388     X=X_OLD+ALPHA*(U_OLD+Q)
1389  !-------------------------------------------------------
1390  ! Stores the matrix in compressed spars row and multiplies by
1391  ! vector
1392  !-------------------------------------------------------
1393     !CALL CSR_MAT_V_PRODUCT      (N,NNZERO,AA_AUX,JA_AUX,IA_AUX,U_OLD+Q,
            AUQ)
1394  !-------------------------------------------------------
1395  ! Stores the matrix in modified spars row and multiplies by
1396  ! vector
1397  !-------------------------------------------------------
1398     !CALL MSR_MAT_V_PRODUCT (TL,N,SA,IJA,U_OLD+Q,AUQ)
1399  !-------------------------------------------------------
1400  ! Stores the matrix in Ellpack/Itpack format and multiplies by
1401  ! vector
1402  !-------------------------------------------------------
1403     !CALL ELLPACK_ITPACK_MAT_V_PRODUCT (N,U_OLD+Q,AUQ,N,IDIAG,COEF,JCOEF
            )
1404  !-------------------------------------------------------
1405  ! Stores the matrix in Diagonal format and multiplies by
1406  ! vector
1407  !-------------------------------------------------------
1408     CALL DIAGONAL_MAT_V_PRODUCT (N,U_OLD+Q,AUQ,DIAG,N,IDIAG,IOFF)
1409  !-------------------------------------------------------
1410     R=R_OLD-ALPHA*AUQ
1411     DO K=1,N
1412        S=S+R(K)**2
1413     END DO
1414     NORM=SQRT(S)/N
```

```
1415 !-------------------------------------------------------
1416    DO K=1,N
1417       MN=MN+R(K)*RS(K)
1418    END DO
1419    BETA=MN/M
1420    U=R+BETA*Q
1421    P=U+BETA*(Q+BETA*P_OLD)
1422    P_OLD=P
1423    X_OLD=X
1424    R_OLD=R
1425    U_OLD=U
1426    !PRINT*,NORM
1427    !CALL SUCCESSIVE_SOL (NORM,ITER)
1428 END DO
1429 !-------------------------------------------------------
1430 END SUBROUTINE CGS
1431 !*****************************************************
1432 !    Bi-Conjugate Gradien Stabilized Method
1433 !*****************************************************
1434 SUBROUTINE BCGSTAB (N,NNZERO,AA_AUX,JA_AUX,IA_AUX,X_OLD,R_OLD,X)
1435 !-------------------------------------------------------
1436 USE PARAM, ONLY                         : TOL
1437 IMPLICIT NONE
1438 INTEGER                                  :: K,ITER,N,NNZERO,TL,IDIAG,IERR
1439 REAL(8),DIMENSION(NNZERO)                :: AA_AUX,AAR
1440 INTEGER,DIMENSION(NNZERO)                :: JA_AUX,JAR
1441 INTEGER,DIMENSION(2*N-1)                 :: IND
1442 INTEGER,DIMENSION(N+1)                   :: IA_AUX,IAR
1443 REAL(8),ALLOCATABLE                      :: SA(:),COEF(:,:),DIAG(:,:)
1444 INTEGER,ALLOCATABLE                      :: IJA(:),JCOEF(:,:),IOFF(:)
1445 REAL(8),DIMENSION(N)                     :: X_OLD,X,P_OLD,P,R_OLD,R,AP,AQQ,RS
        ,QQ
1446 REAL(8)                                  :: NORM,S,ALPHA,BETA,M,MS,MMS,MM,MN,
        OM
1447 !-------------------------------------------------------
1448 CALL TOTAL_LENGTH            (N,IA_AUX,TL)
1449 ALLOCATE                    (SA(TL),IJA(TL))
1450 CALL CSRMSR                 (TL,N,NNZERO,AA_AUX,JA_AUX,IA_AUX,SA,IJA)
1451 CALL INFDIA                 (N,NNZERO,JA_AUX,IA_AUX,IND,IDIAG)
1452 ALLOCATE                    (COEF(N,IDIAG),JCOEF(N,IDIAG))
1453 ALLOCATE                    (IOFF(IDIAG),DIAG(N,IDIAG))
1454 CALL CSRELL                 (N,NNZERO,AA_AUX,JA_AUX,IA_AUX,N,COEF,JCOEF,
        N,IDIAG,IERR)
1455 CALL CSRDIA                 (N,NNZERO,IDIAG,10,AA_AUX,JA_AUX,IA_AUX,N,
        DIAG,IOFF,AAR,JAR,IAR,IND)
1456 !-------------------------------------------------------
1457 RS=R_OLD
1458 P_OLD=R_OLD
1459 NORM=1.D0
1460 ITER=0
1461 DO WHILE (NORM.GT.TOL)
1462    ITER=ITER+1
1463    !PRINT*,ITER
1464    S=0.D0
1465    M=0.D0
1466    MS=0.D0
1467    MMS=0.D0
1468    MM=0.D0
1469    MN=0.D0
1470 !-------------------------------------------------------
1471 ! Stores the matrix in compressed spars row and multiplies by
1472 ! vector
1473 !-------------------------------------------------------
1474    !CALL CSR_MAT_V_PRODUCT    (N,NNZERO,AA_AUX,JA_AUX,IA_AUX,P_OLD,AP)
1475 !-------------------------------------------------------
1476 ! Stores the matrix in modified spars row and multiplies by
1477 ! vector
```

```
1478  !-------------------------------------------------------
1479      !CALL MSR_MAT_V_PRODUCT (TL,N,SA,IJA,P_OLD,AP)
1480  !-------------------------------------------------------
1481  ! Stores the matrix in Ellpack/Itpack format and multiplies by
1482  ! vector
1483  !-------------------------------------------------------
1484      !CALL ELLPACK_ITPACK_MAT_V_PRODUCT (N,P_OLD,AP,N,IDIAG,COEF,JCOEF)
1485  !-------------------------------------------------------
1486  ! Stores the matrix in Diagonal format and multiplies by
1487  ! vector
1488  !-------------------------------------------------------
1489      CALL DIAGONAL_MAT_V_PRODUCT (N,P_OLD,AP,DIAG,N,IDIAG,IOFF)
1490  !-------------------------------------------------------
1491      DO K=1,N
1492         M=M+R_OLD(K)*RS(K)
1493         MM=MM+AP(K)*RS(K)
1494      END DO
1495      ALPHA=M/MM
1496      QQ=R_OLD-ALPHA*AP
1497  !-------------------------------------------------------
1498  ! Stores the matrix in compressed spars row and multiplies by
1499  ! vector
1500  !-------------------------------------------------------
1501      !CALL CSR_MAT_V_PRODUCT     (N,NNZERO,AA_AUX,JA_AUX,IA_AUX,QQ,AQQ)
1502  !-------------------------------------------------------
1503  ! Stores the matrix in modified spars row and multiplies by
1504  ! vector
1505  !-------------------------------------------------------
1506      !CALL MSR_MAT_V_PRODUCT (TL,N,SA,IJA,QQ,AQQ)
1507  !-------------------------------------------------------
1508  ! Stores the matrix in Ellpack/Itpack format and multiplies by
1509  ! vector
1510  !-------------------------------------------------------
1511      !CALL ELLPACK_ITPACK_MAT_V_PRODUCT (N,QQ,AQQ,N,IDIAG,COEF,JCOEF)
1512  !-------------------------------------------------------
1513  ! Stores the matrix in Diagonal format and multiplies by
1514  ! vector
1515  !-------------------------------------------------------
1516      CALL DIAGONAL_MAT_V_PRODUCT (N,QQ,AQQ,DIAG,N,IDIAG,IOFF)
1517  !-------------------------------------------------------
1518      DO K=1,N
1519         MS=MS+AQQ(K)*QQ(K)
1520         MMS=MMS+AQQ(K)*AQQ(K)
1521      END DO
1522      OM=MS/MMS
1523      X=X_OLD+ALPHA*P_OLD+OM*QQ
1524      R=QQ-OM*AQQ
1525      DO K=1,N
1526         S=S+R(K)**2
1527      END DO
1528      NORM=SQRT(S)/N
1529  !-------------------------------------------------------
1530      DO K=1,N
1531         MN=MN+R(K)*RS(K)
1532      END DO
1533      BETA=(MN/M)*(ALPHA/OM)
1534      P=R+BETA*(P_OLD-OM*AP)
1535      P_OLD=P
1536      X_OLD=X
1537      R_OLD=R
1538      !PRINT*,NORM
1539      !CALL SUCCESSIVE_SOL (NORM,ITER)
1540  END DO
1541  !-------------------------------------------------------
1542  END SUBROUTINE BCGSTAB
1543  !************************************************************
1544  !!ILU(0)-Preconditioned Bi-Conjugate Gradient Stabilized Method
```

```
1545  !*****************************************************************
1546  SUBROUTINE ILU0_PBCGSTAB (N,NNZERO,AA_AUX,JA_AUX,IA_AUX,X_OLD,R_OLD,X)
1547  !-----------------------------------------------------
1548  USE PARAM, ONLY                    : TOL
1549  IMPLICIT NONE
1550  INTEGER                            :: ITER,N,NNZERO,TL,ICODE,IDIAG,
          IERR
1551  REAL(8),DIMENSION(NNZERO)          :: AA_AUX,AAR,LUVAL
1552  INTEGER,DIMENSION(NNZERO)          :: JA_AUX,JAR
1553  INTEGER,DIMENSION(2*N-1)           :: IND
1554  INTEGER,DIMENSION(N+1)             :: IA_AUX,IAR
1555  REAL(8),ALLOCATABLE                :: SA(:),COEF(:,:),DIAG(:,:)
1556  INTEGER,ALLOCATABLE                :: IJA(:),JCOEF(:,:),IOFF(:)
1557  REAL(8),DIMENSION(N)               :: X_OLD,X,P_OLD,P,R_OLD,R,RS,SR,
          SRHAT,T,V,V_OLD,PHAT
1558  REAL(8)                            :: NORM,OM,OM_OLD,ALPHA,ALPHA_OLD,
          BETA,RHO,RHO_OLD
1559  INTEGER,DIMENSION(N)               :: UPTR,IW
1560  !-----------------------------------------------------
1561  CALL TOTAL_LENGTH          (N,IA_AUX,TL)
1562  ALLOCATE                   (SA(TL),IJA(TL))
1563  CALL CSRMSR                (TL,N,NNZERO,AA_AUX,JA_AUX,IA_AUX,SA,IJA)
1564  CALL INFDIA                (N,NNZERO,JA_AUX,IA_AUX,IND,IDIAG)
1565  ALLOCATE                   (COEF(N,IDIAG),JCOEF(N,IDIAG))
1566  ALLOCATE                   (IOFF(IDIAG),DIAG(N,IDIAG))
1567  CALL CSRELL                (N,NNZERO,AA_AUX,JA_AUX,IA_AUX,N,COEF,JCOEF
          ,N,IDIAG,IERR)
1568  CALL CSRDIA                (N,NNZERO,IDIAG,10,AA_AUX,JA_AUX,IA_AUX,N,
          DIAG,IOFF,AAR,JAR,IAR,IND)
1569  !-----------------------------------------------------
1570  CALL ILU0     (N,NNZERO,AA_AUX,JA_AUX,IA_AUX,LUVAL,UPTR,IW,ICODE)
1571  !-----------------------------------------------------
1572  RS=R_OLD
1573  NORM=1.D0
1574  ITER=0
1575  DO WHILE (NORM.GT.TOL)
1576      ITER=ITER+1
1577      !PRINT*,ITER
1578      RHO=DOT_PRODUCT(RS,R_OLD)
1579  !-----------------------------------------------------
1580      IF (RHO.EQ.0.D0) PRINT*, 'ILU0_PBiconjugate gradient stabilized
          method fails'
1581  !-----------------------------------------------------
1582      IF (ITER.EQ.1) THEN
1583          P=R_OLD
1584      ELSE
1585          BETA=(RHO/RHO_OLD)*(ALPHA_OLD/OM_OLD)
1586          P=R_OLD+BETA*(P_OLD-OM_OLD*V_OLD)
1587      END IF
1588  !-----------------------------------------------------
1589      CALL LUSOL   (N,NNZERO,P,PHAT,LUVAL,JA_AUX,IA_AUX,UPTR)
1590  !-----------------------------------------------------
1591  ! Stores the matrix in compressed spars row and multiplies by
1592  ! vector
1593  !-----------------------------------------------------
1594      CALL CSR_MAT_V_PRODUCT      (N,NNZERO,AA_AUX,JA_AUX,IA_AUX,PHAT,V)
1595  !-----------------------------------------------------
1596  ! Stores the matrix in modified spars row and multiplies by
1597  ! vector
1598  !-----------------------------------------------------
1599  ! CALL MSR_MAT_V_PRODUCT (TL,N,SA,IJA,PHAT,V)
1600  !-----------------------------------------------------
1601  ! Stores the matrix in Ellpack/Itpack format and multiplies by
1602  ! vector
1603  !-----------------------------------------------------
1604  ! CALL ELLPACK_ITPACK_MAT_V_PRODUCT (N,PHAT,V,N,IDIAG,COEF,JCOEF)
1605  !-----------------------------------------------------
```

```
1606  ! Stores the matrix in Diagonal format and multiplies by
1607  ! vector
1608  !---------------------------------------------------
1609  ! CALL DIAGONAL_MAT_V_PRODUCT (N,PHAT,V,DIAG,N,IDIAG,IOFF)
1610  !---------------------------------------------------
1611      ALPHA=RHO/DOT_PRODUCT(RS,V)
1612      SR=R_OLD-ALPHA*V
1613  !---------------------------------------------------
1614      CALL LUSOL   (N,NNZERO,SR,SRHAT,LUVAL,JA_AUX,IA_AUX,UPTR)
1615  !---------------------------------------------------
1616  ! Stores the matrix in compressed spars row and multiplies by
1617  ! vector
1618  !---------------------------------------------------
1619      CALL CSR_MAT_V_PRODUCT     (N,NNZERO,AA_AUX,JA_AUX,IA_AUX,SRHAT,T)
1620  !---------------------------------------------------
1621  ! Stores the matrix in modified spars row and multiplies by
1622  ! vector
1623  !---------------------------------------------------
1624  ! CALL MSR_MAT_V_PRODUCT (TL,N,SA,IJA,SRHAT,T)
1625  !---------------------------------------------------
1626  ! Stores the matrix in Ellpack/Itpack format and multiplies by
1627  ! vector
1628  !---------------------------------------------------
1629  ! CALL ELLPACK_ITPACK_MAT_V_PRODUCT (N,SRHAT,T,N,IDIAG,COEF,JCOEF)
1630  !---------------------------------------------------
1631  ! Stores the matrix in Diagonal format and multiplies by
1632  ! vector
1633  !---------------------------------------------------
1634  ! CALL DIAGONAL_MAT_V_PRODUCT (N,SRHAT,T,DIAG,N,IDIAG,IOFF)
1635  !---------------------------------------------------
1636      OM=DOT_PRODUCT(T,SR)/DOT_PRODUCT(T,T)
1637      X= X_OLD+ALPHA*PHAT+OM*SRHAT
1638      R=SR-OM*T
1639      NORM=DSQRT(DOT_PRODUCT(R,R))/N
1640  !---------------------------------------------------
1641      X_OLD=X
1642      R_OLD=R
1643      RHO_OLD=RHO
1644      P_OLD=P
1645      OM_OLD=OM
1646      V_OLD=V
1647      ALPHA_OLD=ALPHA
1648      !PRINT*,NORM
1649      !CALL SUCCESSIVE_SOL (NORM,ITER)
1650  END DO
1651  !---------------------------------------------------
1652  END SUBROUTINE ILU0_PBCGSTAB
1653  !*************************************************
1654  !ILU(0)-Preconditioned Conjugate Gradient Squared Method
1655  !*************************************************
1656  SUBROUTINE ILU0_PCGS (N,NNZERO,AA_AUX,JA_AUX,IA_AUX,X_OLD,R_OLD,X)
1657  !---------------------------------------------------
1658  USE PARAM, ONLY                  : TOL
1659  IMPLICIT NONE
1660  INTEGER                          :: ITER,N,NNZERO,TL,ICODE,IDIAG,
         IERR
1661  REAL(8),DIMENSION(NNZERO)        :: AA_AUX,AAR,LUVAL
1662  INTEGER,DIMENSION(NNZERO)        :: JA_AUX,JAR
1663  INTEGER,DIMENSION(2*N-1)         :: IND
1664  INTEGER,DIMENSION(N+1)           :: IA_AUX,IAR
1665  REAL(8),ALLOCATABLE              :: SA(:),COEF(:,:),DIAG(:,:)
1666  INTEGER,ALLOCATABLE             :: IJA(:),JCOEF(:,:),IOFF(:)
1667  REAL(8),DIMENSION(N)            :: X_OLD,X,P_OLD,P,R_OLD,R,RS,Q,
         Q_OLD,UHAT,V,PHAT,U,QHAT
1668  REAL(8)                         :: NORM,ALPHA,BETA,RHO,RHO_OLD
1669  INTEGER,DIMENSION(N)            :: UPTR,IW
1670  !---------------------------------------------------
```

```
1671    CALL TOTAL_LENGTH        (N,IA_AUX,TL)
1672    ALLOCATE                 (SA(TL),IJA(TL))
1673    CALL CSRMSR              (TL,N,NNZERO,AA_AUX,JA_AUX,IA_AUX,SA,IJA)
1674    CALL INFDIA              (N,NNZERO,JA_AUX,IA_AUX,IND,IDIAG)
1675    ALLOCATE                 (COEF(N,IDIAG),JCOEF(N,IDIAG))
1676    ALLOCATE                 (IOFF(IDIAG),DIAG(N,IDIAG))
1677    CALL CSRELL              (N,NNZERO,AA_AUX,JA_AUX,IA_AUX,N,COEF,JCOEF
           ,N,IDIAG,IERR)
1678    CALL CSRDIA              (N,NNZERO,IDIAG,10,AA_AUX,JA_AUX,IA_AUX,N,
           DIAG,IOFF,AAR,JAR,IAR,IND)
1679 !----------------------------------------------------
1680    CALL ILU0    (N,NNZERO,AA_AUX,JA_AUX,IA_AUX,LUVAL,UPTR,IW,ICODE)
1681 !----------------------------------------------------
1682    RS=R_OLD
1683    NORM=1.D0
1684    ITER=0
1685    DO WHILE (NORM.GT.TOL)
1686       ITER=ITER+1
1687       !PRINT*,ITER
1688       RHO=DOT_PRODUCT(RS,R_OLD)
1689 !----------------------------------------------------
1690       IF (RHO.EQ.0.D0) PRINT*, 'conjugate gradient squared method fails'
1691 !----------------------------------------------------
1692       IF (ITER.EQ.1) THEN
1693          U=R_OLD
1694          P=U
1695       ELSE
1696          BETA=RHO/RHO_OLD
1697          U=R_OLD+BETA*Q_OLD
1698          P=U+BETA*(Q_OLD+BETA*P_OLD)
1699       ENDIF
1700 !----------------------------------------------------
1701       CALL LUSOL   (N,NNZERO,P,PHAT,LUVAL,JA_AUX,IA_AUX,UPTR)
1702 !----------------------------------------------------
1703 ! Stores the matrix in compressed spars row and multiplies by
1704 ! vector
1705 !----------------------------------------------------
1706       CALL CSR_MAT_V_PRODUCT (N,NNZERO,AA_AUX,JA_AUX,IA_AUX,PHAT,V)
1707 !----------------------------------------------------
1708 ! Stores the matrix in modified spars row and multiplies by
1709 ! vector
1710 !----------------------------------------------------
1711 ! CALL MSR_MAT_V_PRODUCT (TL,N,SA,IJA,PHAT,V)
1712 !----------------------------------------------------
1713 ! Stores the matrix in Ellpack/Itpack format and multiplies by
1714 ! vector
1715 !----------------------------------------------------
1716 ! CALL ELLPACK_ITPACK_MAT_V_PRODUCT (N,PHAT,V,N,IDIAG,COEF,JCOEF)
1717 !----------------------------------------------------
1718 ! ! Stores the matrix in Diagonal format and multiplies by
1719 ! vector
1720 !----------------------------------------------------
1721 ! CALL DIAGONAL_MAT_V_PRODUCT (N,PHAT,V,DIAG,N,IDIAG,IOFF)
1722 !----------------------------------------------------
1723       ALPHA=RHO/(DOT_PRODUCT(RS,V))
1724       Q=U-ALPHA*V
1725       CALL LUSOL   (N,NNZERO,U+Q,UHAT,LUVAL,JA_AUX,IA_AUX,UPTR)
1726       X=X_OLD+ALPHA*UHAT
1727 !----------------------------------------------------
1728 ! Stores the matrix in compressed spars row and multiplies by
1729 ! vector
1730 !----------------------------------------------------
1731       CALL CSR_MAT_V_PRODUCT     (N,NNZERO,AA_AUX,JA_AUX,IA_AUX,UHAT,QHAT
           )
1732 !----------------------------------------------------
1733 ! Stores the matrix in modified spars row and multiplies by
1734 ! vector
```

```
1735  !------------------------------------------------------
1736  ! CALL MSR_MAT_V_PRODUCT (TL,N,SA,IJA,UHAT,QHAT)
1737  !------------------------------------------------------
1738  ! Stores the matrix in Ellpack/Itpack format and multiplies by
1739  ! vector
1740  !------------------------------------------------------
1741  ! CALL ELLPACK_ITPACK_MAT_V_PRODUCT (N,UHAT,QHAT,N,IDIAG,COEF,JCOEF)
1742  !------------------------------------------------------
1743  ! ! Stores the matrix in Diagonal format and multiplies by
1744  ! vector
1745  !------------------------------------------------------
1746  ! CALL DIAGONAL_MAT_V_PRODUCT (N,UHAT,QHAT,DIAG,N,IDIAG,IOFF)
1747  !------------------------------------------------------
1748       R=R_OLD-ALPHA*QHAT
1749       NORM=DSQRT(DOT_PRODUCT(R,R))/N
1750  !------------------------------------------------------
1751       P_OLD=P
1752       X_OLD=X
1753       R_OLD=R
1754       Q_OLD=Q
1755       RHO_OLD=RHO
1756       !PRINT*,NORM
1757       !CALL SUCCESSIVE_SOL (NORM,ITER)
1758  END DO
1759  !------------------------------------------------------
1760  END SUBROUTINE ILU0_PCGS
1761  !***************************************************
1762  !    Conjugate Gradien Method Inner Iteration
1763  !***************************************************
1764  SUBROUTINE CG_INNER (MD,NNZERO,INNER_ITER,AA_AUX,JA_AUX,IA_AUX,X_OLD,B,
             X)
1765  !------------------------------------------------------
1766  USE PARAM, ONLY                      : TOL
1767  IMPLICIT NONE
1768  INTEGER                              :: K,TL,MD,NNZERO,IDIAG,IERR,I,
             INNER_ITER
1769  REAL(8),DIMENSION(MD)                :: X_OLD,X,P_OLD,P,R_OLD,R,AP,B
1770  INTEGER,DIMENSION(MD+1)              :: IA_AUX,IAR
1771  INTEGER,DIMENSION(2*MD-1)            :: IND
1772  REAL(8)                              :: S,ALPHA,BETA,M,MM
1773  REAL(8),DIMENSION(NNZERO)            :: AA_AUX,AAR
1774  INTEGER,DIMENSION(NNZERO)            :: JA_AUX,JAR
1775  REAL(8),ALLOCATABLE                  :: SA(:),DIAG(:,:),COEF(:,:)
1776  INTEGER,ALLOCATABLE                  :: IJA(:),IOFF(:),JCOEF(:,:)
1777  !------------------------------------------------------
1778  CALL TOTAL_LENGTH          (MD,IA_AUX,TL)
1779  ALLOCATE                   (SA(TL),IJA(TL))
1780  CALL CSRMSR                (TL,MD,NNZERO,AA_AUX,JA_AUX,IA_AUX,SA,IJA)
1781  CALL INFDIA                (MD,NNZERO,JA_AUX,IA_AUX,IND,IDIAG)
1782  ALLOCATE                   (COEF(MD,IDIAG),JCOEF(MD,IDIAG))
1783  ALLOCATE                   (IOFF(IDIAG),DIAG(MD,IDIAG))
1784  CALL CSRELL                (MD,NNZERO,AA_AUX,JA_AUX,IA_AUX,MD,COEF,
             JCOEF,MD,IDIAG,IERR)
1785  CALL CSRDIA                (MD,NNZERO,IDIAG,10,AA_AUX,JA_AUX,IA_AUX,MD,
             DIAG,IOFF,AAR,JAR,IAR,IND)
1786  !------------------------------------------------------
1787  CALL INITIAL_RESIDUAL (MD,NNZERO,AA_AUX,JA_AUX,IA_AUX,B,X_OLD,R_OLD)
1788  !------------------------------------------------------
1789  P_OLD=R_OLD
1790  DO I=1,INNER_ITER
1791       S=0.D0
1792       M=0.D0
1793       MM=0.D0
1794  !------------------------------------------------------
1795  ! Stores the matrix in compressed spars row and multiplies by
1796  ! vector
1797  !------------------------------------------------------
```

```
1798  !CALL CSR_MAT_V_PRODUCT (MD,NNZERO,AA_AUX,JA_AUX,IA_AUX,P_OLD,AP)
1799  !-------------------------------------------------
1800  ! Stores the matrix in modified spars row and multiplies by
1801  ! vector
1802  !-------------------------------------------------
1803      CALL MSR_MAT_V_PRODUCT (TL,MD,SA,IJA,P_OLD,AP)
1804  !-------------------------------------------------
1805  ! Stores the matrix in Ellpack/Itpack format and multiplies by
1806  ! vector
1807  !-------------------------------------------------
1808  !CALL ELLPACK_ITPACK_MAT_V_PRODUCT (MD,P_OLD,AP,MD,IDIAG,COEF,JCOEF)
1809  !-------------------------------------------------
1810  ! Stores the matrix in Diagonal format and multiplies by
1811  ! vector
1812  !-------------------------------------------------
1813  !CALL  DIAGONAL_MAT_V_PRODUCT (MD,P_OLD,AP,DIAG,MD,IDIAG,IOFF)
1814  !-------------------------------------------------
1815      DO K=1,MD
1816          M=M+R_OLD(K)**2
1817          MM=MM+AP(K)*P_OLD(K)
1818      END DO
1819      ALPHA=M/MM
1820      X=X_OLD+ALPHA*P_OLD
1821      R=R_OLD-ALPHA*AP
1822      DO K=1,MD
1823          S=S+R(K)**2
1824      END DO
1825      BETA=S/M
1826      P=R+BETA*P_OLD
1827      P_OLD=P
1828      X_OLD=X
1829      R_OLD=R
1830  END DO
1831  !-------------------------------------------------
1832  END SUBROUTINE CG_INNER
1833  !*****************************************************************
1834  !ILU0 Preconditioned Conjugate Gradient Method Inner Iteration
1835  !*****************************************************************
1836  SUBROUTINE ILU0_PCG_INNER (N,NNZERO,INNER_ITER,AA_AUX,JA_AUX,IA_AUX,
           X_OLD,B,X)
1837  !-------------------------------------------------
1838  USE PARAM, ONLY                      : TOL
1839  IMPLICIT NONE
1840  INTEGER                              :: K,I,TL,N,NNZERO,ICODE,IDIAG,IERR,
           INNER_ITER
1841  REAL(8),DIMENSION(N)                 :: X_OLD,X,P_OLD,P,R_OLD,R,AP,Z_OLD,
           Z,B
1842  INTEGER,DIMENSION(N+1)               :: IA_AUX,IAR
1843  INTEGER,DIMENSION(2*N-1)             :: IND
1844  REAL(8)                              :: S,ALPHA,BETA,M,MM
1845  REAL(8),DIMENSION(NNZERO)            :: AA_AUX,AAR,LUVAL
1846  INTEGER,DIMENSION(NNZERO)            :: JA_AUX,JAR
1847  REAL(8),ALLOCATABLE                  :: SA(:),COEF(:,:),DIAG(:,:)
1848  INTEGER,ALLOCATABLE                  :: IJA(:),JCOEF(:,:),IOFF(:)
1849  INTEGER,DIMENSION(N)                 :: UPTR,IW
1850  REAL(8)                              :: SN
1851  !-------------------------------------------------
1852  CALL TOTAL_LENGTH             (N,IA_AUX,TL)
1853  ALLOCATE                      (SA(TL),IJA(TL))
1854  CALL CSRMSR                   (TL,N,NNZERO,AA_AUX,JA_AUX,IA_AUX,SA,IJA)
1855  CALL INFDIA                   (N,NNZERO,JA_AUX,IA_AUX,IND,IDIAG)
1856  ALLOCATE                      (COEF(N,IDIAG),JCOEF(N,IDIAG))
1857  ALLOCATE                      (IOFF(IDIAG),DIAG(N,IDIAG))
1858  CALL CSRELL                   (N,NNZERO,AA_AUX,JA_AUX,IA_AUX,N,COEF,JCOEF,
           N,IDIAG,IERR)
1859  CALL CSRDIA                   (N,NNZERO,IDIAG,10,AA_AUX,JA_AUX,IA_AUX,N,
           DIAG,IOFF,AAR,JAR,IAR,IND)
```

```
1860  !----------------------------------------------------
1861  CALL INITIAL_RESIDUAL (N,NNZERO,AA_AUX,JA_AUX,IA_AUX,B,X_OLD,R_OLD)
1862  !----------------------------------------------------
1863  CALL ILU0     (N,NNZERO,AA_AUX,JA_AUX,IA_AUX,LUVAL,UPTR,IW,ICODE)
1864  CALL LUSOL    (N,NNZERO,R_OLD,Z_OLD,LUVAL,JA_AUX,IA_AUX,UPTR)
1865  !----------------------------------------------------
1866  P_OLD=Z_OLD
1867  DO I=1,INNER_ITER
1868     SN=0.D0
1869     S=0.D0
1870     M=0.D0
1871     MM=0.D0
1872  !----------------------------------------------------
1873  ! Stores the matrix in compressed spars row and multiplies by
1874  ! vector
1875  !----------------------------------------------------
1876  !CALL CSR_MAT_V_PRODUCT      (N,NNZERO,AA_AUX,JA_AUX,IA_AUX,P_OLD,AP)
1877  !----------------------------------------------------
1878  ! Stores the matrix in modified spars row and multiplies by
1879  ! vector
1880  !----------------------------------------------------
1881  !CALL MSR_MAT_V_PRODUCT (TL,N,SA,IJA,P_OLD,AP)
1882  !----------------------------------------------------
1883  ! Stores the matrix in Ellpack/Itpack format and multiplies by
1884  ! vector
1885  !----------------------------------------------------
1886  !CALL ELLPACK_ITPACK_MAT_V_PRODUCT (N,P_OLD,AP,N,IDIAG,COEF,JCOEF)
1887  !----------------------------------------------------
1888  ! Stores the matrix in Diagonal format and multiplies by
1889  ! vector
1890  !----------------------------------------------------
1891     CALL DIAGONAL_MAT_V_PRODUCT (N,P_OLD,AP,DIAG,N,IDIAG,IOFF)
1892  !----------------------------------------------------
1893     DO K=1,N
1894        M=M+R_OLD(K)*Z_OLD(K)
1895        MM=MM+AP(K)*P_OLD(K)
1896     END DO
1897     ALPHA=M/MM
1898     X=X_OLD+ALPHA*P_OLD
1899     R=R_OLD-ALPHA*AP
1900  !----------------------------------------------------
1901     CALL LUSOL    (N,NNZERO,R,Z,LUVAL,JA_AUX,IA_AUX,UPTR)
1902  !----------------------------------------------------
1903     DO K=1,N
1904        S=S+R(K)*Z(K)
1905     END DO
1906     BETA=S/M
1907     P=Z+BETA*P_OLD
1908     P_OLD=P
1909     X_OLD=X
1910     R_OLD=R
1911     Z_OLD=Z
1912  END DO
1913  !----------------------------------------------------
1914  END SUBROUTINE ILU0_PCG_INNER
1915  !****************************************************
1916  !ILU0 Preconditioned Bi-Conjugate Gradient
1917  !Stabilized Method Inner Iteration
1918  !****************************************************
1919  SUBROUTINE ILU0_PBCGSTAB_INNER (N,NNZERO,INNER_ITER,AA_AUX,JA_AUX,
         IA_AUX,X_OLD,B,X)
1920  !----------------------------------------------------
1921  USE PARAM, ONLY              : TOL
1922  IMPLICIT NONE
1923  INTEGER                      :: I,N,NNZERO,TL,ICODE,IDIAG,IERR,
         INNER_ITER
1924  REAL(8),DIMENSION(NNZERO)    :: AA_AUX,AAR,LUVAL
```

```
1925  INTEGER,DIMENSION(NNZERO)               :: JA_AUX,JAR
1926  INTEGER,DIMENSION(2*N-1)                 :: IND
1927  INTEGER,DIMENSION(N+1)                   :: IA_AUX,IAR
1928  REAL(8),ALLOCATABLE                      :: SA(:),COEF(:,:),DIAG(:,:)
1929  INTEGER,ALLOCATABLE                      :: IJA(:),JCOEF(:,:),IOFF(:)
1930  REAL(8),DIMENSION(N)                     :: X_OLD,X,P_OLD,P,R_OLD,R,RS,SR,
          SRHAT,T,V,V_OLD,PHAT,B
1931  REAL(8)                                  :: OM,OM_OLD,ALPHA,ALPHA_OLD,BETA,
          RHO,RHO_OLD
1932  INTEGER,DIMENSION(N)                     :: UPTR,IW
1933  !-------------------------------------------------------
1934  CALL TOTAL_LENGTH             (N,IA_AUX,TL)
1935  ALLOCATE                      (SA(TL),IJA(TL))
1936  CALL CSRMSR                   (TL,N,NNZERO,AA_AUX,JA_AUX,IA_AUX,SA,IJA)
1937  CALL INFDIA                   (N,NNZERO,JA_AUX,IA_AUX,IND,IDIAG)
1938  ALLOCATE                      (COEF(N,IDIAG),JCOEF(N,IDIAG))
1939  ALLOCATE                      (IOFF(IDIAG),DIAG(N,IDIAG))
1940  CALL CSRELL                   (N,NNZERO,AA_AUX,JA_AUX,IA_AUX,N,COEF,JCOEF,
          N,IDIAG,IERR)
1941  CALL CSRDIA                   (N,NNZERO,IDIAG,10,AA_AUX,JA_AUX,IA_AUX,N,
          DIAG,IOFF,AAR,JAR,IAR,IND)
1942  !-------------------------------------------------------
1943  CALL INITIAL_RESIDUAL (N,NNZERO,AA_AUX,JA_AUX,IA_AUX,B,X_OLD,R_OLD)
1944  !-------------------------------------------------------
1945  CALL ILU0    (N,NNZERO,AA_AUX,JA_AUX,IA_AUX,LUVAL,UPTR,IW,ICODE)
1946  !-------------------------------------------------------
1947  RS=R_OLD
1948  DO I=1,INNER_ITER
1949     RHO=DOT_PRODUCT(RS,R_OLD)
1950  !-------------------------------------------------------
1951     IF (RHO.EQ.0.D0) PRINT*, 'Biconjugate gradient stabilized method
          fails'
1952  !-------------------------------------------------------
1953     IF (I.EQ.1) THEN
1954        P=R_OLD
1955     ELSE
1956        BETA=(RHO/RHO_OLD)*(ALPHA_OLD/OM_OLD)
1957        P=R_OLD+BETA*(P_OLD-OM_OLD*V_OLD)
1958     END IF
1959  !-------------------------------------------------------
1960     CALL LUSOL   (N,NNZERO,P,PHAT,LUVAL,JA_AUX,IA_AUX,UPTR)
1961  !-------------------------------------------------------
1962  !Stores the matrix in compressed spars row and multiplies by
1963  !vector
1964  !-------------------------------------------------------
1965  !CALL CSR_MAT_V_PRODUCT        (N,NNZERO,AA_AUX,JA_AUX,IA_AUX,PHAT,V)
1966  !-------------------------------------------------------
1967  ! Stores the matrix in modified spars row and multiplies by
1968  ! vector
1969  !-------------------------------------------------------
1970     CALL MSR_MAT_V_PRODUCT (TL,N,SA,IJA,PHAT,V)
1971  !-------------------------------------------------------
1972  !Stores the matrix in Ellpack/Itpack format and multiplies by
1973  !vector
1974  !-------------------------------------------------------
1975  !CALL ELLPACK_ITPACK_MAT_V_PRODUCT (N,PHAT,V,N,IDIAG,COEF,JCOEF)
1976  !-------------------------------------------------------
1977  !Stores the matrix in Diagonal format and multiplies by
1978  !vector
1979  !-------------------------------------------------------
1980  !CALL DIAGONAL_MAT_V_PRODUCT (N,PHAT,V,DIAG,N,IDIAG,IOFF)
1981  !-------------------------------------------------------
1982     ALPHA=RHO/DOT_PRODUCT(RS,V)
1983     SR=R_OLD-ALPHA*V
1984  !-------------------------------------------------------
1985     CALL LUSOL   (N,NNZERO,SR,SRHAT,LUVAL,JA_AUX,IA_AUX,UPTR)
1986  !-------------------------------------------------------
```

```
1987  !Stores the matrix in compressed spars row and multiplies by
1988  !vector
1989  !-------------------------------------------------
1990  !CALL CSR_MAT_V_PRODUCT        (N,NNZERO,AA_AUX,JA_AUX,IA_AUX,SRHAT,T)
1991  !-------------------------------------------------
1992  ! Stores the matrix in modified spars row and multiplies by
1993  ! vector
1994  !-------------------------------------------------
1995     CALL MSR_MAT_V_PRODUCT (TL,N,SA,IJA,SRHAT,T)
1996  !-------------------------------------------------
1997  !Stores the matrix in Ellpack/Itpack format and multiplies by
1998  !vector
1999  !-------------------------------------------------
2000  !CALL ELLPACK_ITPACK_MAT_V_PRODUCT (N,SRHAT,T,N,IDIAG,COEF,JCOEF)
2001  !-------------------------------------------------
2002  !Stores the matrix in Diagonal format and multiplies by
2003  !vector
2004  !-------------------------------------------------
2005  !CALL DIAGONAL_MAT_V_PRODUCT (N,SRHAT,T,DIAG,N,IDIAG,IOFF)
2006  !-------------------------------------------------
2007     OM=DOT_PRODUCT(T,SR)/DOT_PRODUCT(T,T)
2008     X= X_OLD+ALPHA*PHAT+OM*SRHAT
2009     R=SR-OM*T
2010     X_OLD=X
2011     R_OLD=R
2012     RHO_OLD=RHO
2013     P_OLD=P
2014     OM_OLD=OM
2015     V_OLD=V
2016     ALPHA_OLD=ALPHA
2017  END DO
2018  !-------------------------------------------------
2019  END SUBROUTINE ILU0_PBCGSTAB_INNER
2020  !*************************************************
2021  !Conjugate Gradien Squared Method Inner Iteration
2022  !*************************************************
2023  SUBROUTINE CGS_INNER (N,NNZERO,INNER_ITER,AA_AUX,JA_AUX,IA_AUX,X_OLD,B,
         X)
2024  !-------------------------------------------------
2025  USE PARAM, ONLY                   : TOL
2026  IMPLICIT NONE
2027  INTEGER                          :: I,K,N,NNZERO,TL,IDIAG,IERR,
         INNER_ITER
2028  REAL(8),DIMENSION(NNZERO)        :: AA_AUX,AAR
2029  INTEGER,DIMENSION(NNZERO)        :: JA_AUX,JAR
2030  INTEGER,DIMENSION(2*N-1)         :: IND
2031  INTEGER,DIMENSION(N+1)           :: IA_AUX,IAR
2032  REAL(8),ALLOCATABLE              :: SA(:),COEF(:,:),DIAG(:,:)
2033  INTEGER,ALLOCATABLE              :: IJA(:),JCOEF(:,:),IOFF(:)
2034  REAL(8),DIMENSION(N)             :: X_OLD,X,P_OLD,P,R_OLD,U,U_OLD,R,
         AP,AUQ,RS,Q,B
2035  REAL(8)                          :: ALPHA,BETA,M,MM,MN
2036  !-------------------------------------------------
2037  CALL TOTAL_LENGTH           (N,IA_AUX,TL)
2038  ALLOCATE                    (SA(TL),IJA(TL))
2039  CALL CSRMSR                 (TL,N,NNZERO,AA_AUX,JA_AUX,IA_AUX,SA,IJA)
2040  CALL INFDIA                 (N,NNZERO,JA_AUX,IA_AUX,IND,IDIAG)
2041  ALLOCATE                    (COEF(N,IDIAG),JCOEF(N,IDIAG))
2042  ALLOCATE                    (IOFF(IDIAG),DIAG(N,IDIAG))
2043  CALL CSRELL                 (N,NNZERO,AA_AUX,JA_AUX,IA_AUX,N,COEF,JCOEF,
         N,IDIAG,IERR)
2044  CALL CSRDIA                 (N,NNZERO,IDIAG,10,AA_AUX,JA_AUX,IA_AUX,N,
         DIAG,IOFF,AAR,JAR,IAR,IND)
2045  !-------------------------------------------------
2046  CALL INITIAL_RESIDUAL (N,NNZERO,AA_AUX,JA_AUX,IA_AUX,B,X_OLD,R_OLD)
2047  !-------------------------------------------------
2048  RS=R_OLD
```

```
2049  P_OLD=R_OLD
2050  U_OLD=R_OLD
2051  DO I=1,INNER_ITER
2052    M=0.D0
2053    MM=0.D0
2054    MN=0.D0
2055  !--------------------------------------------------
2056  ! Stores the matrix in compressed spars row and multiplies by
2057  ! vector
2058  !--------------------------------------------------
2059  !CALL CSR_MAT_V_PRODUCT (N,NNZERO,AA_AUX,JA_AUX,IA_AUX,P_OLD,AP)
2060  !--------------------------------------------------
2061  ! Stores the matrix in modified spars row and multiplies by
2062  ! vector
2063  !--------------------------------------------------
2064  !CALL MSR_MAT_V_PRODUCT (TL,N,SA,IJA,P_OLD,AP)
2065  !--------------------------------------------------
2066  ! Stores the matrix in Ellpack/Itpack format and multiplies by
2067  ! vector
2068  !--------------------------------------------------
2069  !CALL ELLPACK_ITPACK_MAT_V_PRODUCT (N,P_OLD,AP,N,IDIAG,COEF,JCOEF)
2070  !--------------------------------------------------
2071  ! Stores the matrix in Diagonal format and multiplies by
2072  ! vector
2073  !--------------------------------------------------
2074    CALL DIAGONAL_MAT_V_PRODUCT (N,P_OLD,AP,DIAG,N,IDIAG,IOFF)
2075  !--------------------------------------------------
2076    DO K=1,N
2077      M=M+R_OLD(K)*RS(K)
2078      MM=MM+AP(K)*RS(K)
2079    END DO
2080    ALPHA=M/MM
2081    Q=U_OLD-ALPHA*AP
2082    X=X_OLD+ALPHA*(U_OLD+Q)
2083  !--------------------------------------------------
2084  ! Stores the matrix in compressed spars row and multiplies by
2085  ! vector
2086  !--------------------------------------------------
2087  !CALL CSR_MAT_V_PRODUCT      (N,NNZERO,AA_AUX,JA_AUX,IA_AUX,U_OLD+Q,AUQ)
2088  !--------------------------------------------------
2089  ! Stores the matrix in modified spars row and multiplies by
2090  ! vector
2091  !--------------------------------------------------
2092  !CALL MSR_MAT_V_PRODUCT (TL,N,SA,IJA,U_OLD+Q,AUQ)
2093  !--------------------------------------------------
2094  ! Stores the matrix in Ellpack/Itpack format and multiplies by
2095  ! vector
2096  !--------------------------------------------------
2097  !CALL ELLPACK_ITPACK_MAT_V_PRODUCT (N,U_OLD+Q,AUQ,N,IDIAG,COEF,JCOEF)
2098  !--------------------------------------------------
2099  ! Stores the matrix in Diagonal format and multiplies by
2100  ! vector
2101  !--------------------------------------------------
2102    CALL DIAGONAL_MAT_V_PRODUCT (N,U_OLD+Q,AUQ,DIAG,N,IDIAG,IOFF)
2103  !--------------------------------------------------
2104    R=R_OLD-ALPHA*AUQ
2105    DO K=1,N
2106      MN=MN+R(K)*RS(K)
2107    END DO
2108    BETA=MN/M
2109    U=R+BETA*Q
2110    P=U+BETA*(Q+BETA*P_OLD)
2111    P_OLD=P
2112    X_OLD=X
2113    R_OLD=R
2114    U_OLD=U
2115  END DO
```

```
2116  !-----------------------------------------------------
2117  END SUBROUTINE CGS_INNER
2118  !*****************************************************************
2119  !ILU0 Preconditioned Conjugate Gradient Squared Method Inner Iteration
2120  !*****************************************************************
2121  SUBROUTINE ILU0_PCGS_INNER  (N,NNZERO,INNER_ITER,AA_AUX,JA_AUX,IA_AUX,
            X_OLD,B,X)
2122  !-----------------------------------------------------
2123  USE PARAM, ONLY                 : TOL
2124  IMPLICIT NONE
2125  INTEGER                         :: I,N,NNZERO,TL,ICODE,IDIAG,IERR,
            INNER_ITER
2126  REAL(8),DIMENSION(NNZERO)       :: AA_AUX,AAR,LUVAL
2127  INTEGER,DIMENSION(NNZERO)       :: JA_AUX,JAR
2128  INTEGER,DIMENSION(2*N-1)        :: IND
2129  INTEGER,DIMENSION(N+1)          :: IA_AUX,IAR
2130  REAL(8),ALLOCATABLE             :: SA(:),COEF(:,:),DIAG(:,:)
2131  INTEGER,ALLOCATABLE             :: IJA(:),JCOEF(:,:),IOFF(:)
2132  REAL(8),DIMENSION(N)            :: X_OLD,X,P_OLD,P,R_OLD,R,RS,Q,
            Q_OLD,UHAT,V,PHAT,U,QHAT,B
2133  REAL(8)                         :: ALPHA,BETA,RHO,RHO_OLD
2134  INTEGER,DIMENSION(N)            :: UPTR,IW
2135  !-----------------------------------------------------
2136  CALL TOTAL_LENGTH              (N,IA_AUX,TL)
2137  ALLOCATE                      (SA(TL),IJA(TL))
2138  CALL CSRMSR                   (TL,N,NNZERO,AA_AUX,JA_AUX,IA_AUX,SA,IJA)
2139  CALL INFDIA                   (N,NNZERO,JA_AUX,IA_AUX,IND,IDIAG)
2140  ALLOCATE                      (COEF(N,IDIAG),JCOEF(N,IDIAG))
2141  ALLOCATE                      (IOFF(IDIAG),DIAG(N,IDIAG))
2142  CALL CSRELL                   (N,NNZERO,AA_AUX,JA_AUX,IA_AUX,N,COEF,JCOEF,
            N,IDIAG,IERR)
2143  CALL CSRDIA                   (N,NNZERO,IDIAG,10,AA_AUX,JA_AUX,IA_AUX,N,
            DIAG,IOFF,AAR,JAR,IAR,IND)
2144  !-----------------------------------------------------
2145  CALL INITIAL_RESIDUAL (N,NNZERO,AA_AUX,JA_AUX,IA_AUX,B,X_OLD,R_OLD)
2146  !-----------------------------------------------------
2147  CALL ILU0     (N,NNZERO,AA_AUX,JA_AUX,IA_AUX,LUVAL,UPTR,IW,ICODE)
2148  !-----------------------------------------------------
2149  RS=R_OLD
2150  DO I=1,INNER_ITER
2151     RHO=DOT_PRODUCT(RS,R_OLD)
2152  !-----------------------------------------------------
2153     IF (RHO.EQ.0.D0) PRINT*, 'conjugate gradient squared method fails'
2154  !-----------------------------------------------------
2155     IF (I.EQ.1) THEN
2156        U=R_OLD
2157        P=U
2158     ELSE
2159        BETA=RHO/RHO_OLD
2160        U=R_OLD+BETA*Q_OLD
2161        P=U+BETA*(Q_OLD+BETA*P_OLD)
2162     ENDIF
2163  !-----------------------------------------------------
2164     CALL LUSOL   (N,NNZERO,P,PHAT,LUVAL,JA_AUX,IA_AUX,UPTR)
2165  !-----------------------------------------------------
2166  !Stores the matrix in compressed spars row and multiplies by
2167  !vector
2168  !-----------------------------------------------------
2169  !CALL CSR_MAT_V_PRODUCT     (N,NNZERO,AA_AUX,JA_AUX,IA_AUX,PHAT,V)
2170  !-----------------------------------------------------
2171  !Stores the matrix in modified spars row and multiplies by
2172  !vector
2173  !-----------------------------------------------------
2174     CALL MSR_MAT_V_PRODUCT (TL,N,SA,IJA,PHAT,V)
2175  !-----------------------------------------------------
2176  !Stores the matrix in Ellpack/Itpack format and multiplies by
2177  !vector
```

```
2178  !----------------------------------------------------
2179  !CALL ELLPACK_ITPACK_MAT_V_PRODUCT (N,PHAT,V,N,IDIAG,COEF,JCOEF)
2180  !----------------------------------------------------
2181  !Stores the matrix in Diagonal format and multiplies by
2182  !vector
2183  !----------------------------------------------------
2184  !CALL DIAGONAL_MAT_V_PRODUCT (N,PHAT,V,DIAG,N,IDIAG,IOFF)
2185  !----------------------------------------------------
2186     ALPHA=RHO/(DOT_PRODUCT(RS,V))
2187     Q=U-ALPHA*V
2188     CALL LUSOL   (N,NNZERO,U+Q,UHAT,LUVAL,JA_AUX,IA_AUX,UPTR)
2189     X=X_OLD+ALPHA*UHAT
2190  !----------------------------------------------------
2191  !Stores the matrix in compressed spars row and multiplies by
2192  !vector
2193  !----------------------------------------------------
2194  !CALL CSR_MAT_V_PRODUCT    (N,NNZERO,AA_AUX,JA_AUX,IA_AUX,UHAT,QHAT)
2195  !----------------------------------------------------
2196  !Stores the matrix in modified spars row and multiplies by
2197  !vector
2198  !----------------------------------------------------
2199     CALL MSR_MAT_V_PRODUCT (TL,N,SA,IJA,UHAT,QHAT)
2200  !----------------------------------------------------
2201  !Stores the matrix in Ellpack/Itpack format and multiplies by
2202  !vector
2203  !----------------------------------------------------
2204  !CALL ELLPACK_ITPACK_MAT_V_PRODUCT (N,UHAT,QHAT,N,IDIAG,COEF,JCOEF)
2205  !----------------------------------------------------
2206  !Stores the matrix in Diagonal format and multiplies by
2207  !vector
2208  !----------------------------------------------------
2209  !CALL DIAGONAL_MAT_V_PRODUCT (N,UHAT,QHAT,DIAG,N,IDIAG,IOFF)
2210  !----------------------------------------------------
2211     R=R_OLD-ALPHA*QHAT
2212     P_OLD=P
2213     X_OLD=X
2214     R_OLD=R
2215     Q_OLD=Q
2216     RHO_OLD=RHO
2217  END DO
2218  !----------------------------------------------------
2219  END SUBROUTINE ILU0_PCGS_INNER
2220  !********************************************************
2221  !Bi-Conjugate Gradient Stabilized Method Inner Iteration
2222  !********************************************************
2223  SUBROUTINE BCGSTAB_INNER (N,NNZERO,INNER_ITER,AA_AUX,JA_AUX,IA_AUX,
          X_OLD,B,X)
2224  !----------------------------------------------------
2225  USE PARAM, ONLY              : TOL
2226  IMPLICIT NONE
2227  INTEGER                      :: I,K,N,NNZERO,TL,IDIAG,IERR,
          INNER_ITER
2228  REAL(8),DIMENSION(NNZERO)    :: AA_AUX,AAR
2229  INTEGER,DIMENSION(NNZERO)    :: JA_AUX,JAR
2230  INTEGER,DIMENSION(2*N-1)     :: IND
2231  INTEGER,DIMENSION(N+1)       :: IA_AUX,IAR
2232  REAL(8),ALLOCATABLE          :: SA(:),COEF(:,:),DIAG(:,:)
2233  INTEGER,ALLOCATABLE          :: IJA(:),JCOEF(:,:),IOFF(:)
2234  REAL(8),DIMENSION(N)         :: X_OLD,X,P_OLD,P,R_OLD,R,AP,AQQ,RS
          ,QQ,B
2235  REAL(8)                      :: ALPHA,BETA,M,MS,MMS,MM,MN,OM
2236  !----------------------------------------------------
2237  CALL TOTAL_LENGTH           (N,IA_AUX,TL)
2238  ALLOCATE                    (SA(TL),IJA(TL))
2239  CALL CSRMSR                 (TL,N,NNZERO,AA_AUX,JA_AUX,IA_AUX,SA,IJA)
2240  CALL INFDIA                 (N,NNZERO,JA_AUX,IA_AUX,IND,IDIAG)
2241  ALLOCATE                    (COEF(N,IDIAG),JCOEF(N,IDIAG))
```

```
2242  ALLOCATE                      (IOFF(IDIAG),DIAG(N,IDIAG))
2243  CALL CSRELL                   (N,NNZERO,AA_AUX,JA_AUX,IA_AUX,N,COEF,JCOEF,
          N,IDIAG,IERR)
2244  CALL CSRDIA                   (N,NNZERO,IDIAG,10,AA_AUX,JA_AUX,IA_AUX,N,
          DIAG,IOFF,AAR,JAR,IAR,IND)
2245  !--------------------------------------------------
2246  CALL INITIAL_RESIDUAL  (N,NNZERO,AA_AUX,JA_AUX,IA_AUX,B,X_OLD,R_OLD)
2247  !--------------------------------------------------
2248  RS=R_OLD
2249  P_OLD=R_OLD
2250  DO I=1,INNER_ITER
2251      M=0.D0
2252      MS=0.D0
2253      MMS=0.D0
2254      MM=0.D0
2255      MN=0.D0
2256  !--------------------------------------------------
2257  ! Stores the matrix in compressed spars row and multiplies by
2258  ! vector
2259  !--------------------------------------------------
2260  !CALL CSR_MAT_V_PRODUCT     (N,NNZERO,AA_AUX,JA_AUX,IA_AUX,P_OLD,AP)
2261  !--------------------------------------------------
2262  ! Stores the matrix in modified spars row and multiplies by
2263  ! vector
2264  !--------------------------------------------------
2265      CALL MSR_MAT_V_PRODUCT (TL,N,SA,IJA,P_OLD,AP)
2266  !--------------------------------------------------
2267  ! Stores the matrix in Ellpack/Itpack format and multiplies by
2268  ! vector
2269  !--------------------------------------------------
2270  !CALL ELLPACK_ITPACK_MAT_V_PRODUCT (N,P_OLD,AP,N,IDIAG,COEF,JCOEF)
2271  !--------------------------------------------------
2272  ! Stores the matrix in Diagonal format and multiplies by
2273  ! vector
2274  !--------------------------------------------------
2275  !CALL DIAGONAL_MAT_V_PRODUCT (N,P_OLD,AP,DIAG,N,IDIAG,IOFF)
2276  !--------------------------------------------------
2277      DO K=1,N
2278          M=M+R_OLD(K)*RS(K)
2279          MM=MM+AP(K)*RS(K)
2280      END DO
2281      ALPHA=M/MM
2282      QQ=R_OLD-ALPHA*AP
2283  !--------------------------------------------------
2284  ! Stores the matrix in compressed spars row and multiplies by
2285  ! vector
2286  !--------------------------------------------------
2287  !CALL CSR_MAT_V_PRODUCT     (N,NNZERO,AA_AUX,JA_AUX,IA_AUX,QQ,AQQ)
2288  !--------------------------------------------------
2289  ! Stores the matrix in modified spars row and multiplies by
2290  ! vector
2291  !--------------------------------------------------
2292      CALL MSR_MAT_V_PRODUCT (TL,N,SA,IJA,QQ,AQQ)
2293  !--------------------------------------------------
2294  ! Stores the matrix in Ellpack/Itpack format and multiplies by
2295  ! vector
2296  !--------------------------------------------------
2297  !CALL ELLPACK_ITPACK_MAT_V_PRODUCT (N,QQ,AQQ,N,IDIAG,COEF,JCOEF)
2298  !--------------------------------------------------
2299  ! Stores the matrix in Diagonal format and multiplies by
2300  ! vector
2301  !--------------------------------------------------
2302  !CALL DIAGONAL_MAT_V_PRODUCT (N,QQ,AQQ,DIAG,N,IDIAG,IOFF)
2303  !--------------------------------------------------
2304      DO K=1,N
2305          MS=MS+AQQ(K)*QQ(K)
2306          MMS=MMS+AQQ(K)*AQQ(K)
```

```
2307      END DO
2308      OM=MS/MMS
2309      X=X_OLD+ALPHA*P_OLD+OM*QQ
2310      R=QQ-OM*AQQ
2311      DO K=1,N
2312         MN=MN+R(K)*RS(K)
2313      END DO
2314      BETA=(MN/M)*(ALPHA/OM)
2315      P=R+BETA*(P_OLD-OM*AP)
2316      P_OLD=P
2317      X_OLD=X
2318      R_OLD=R
2319   END DO
2320   !-------------------------------------------------
2321   END SUBROUTINE BCGSTAB_INNER
2322   !**************************************************
2323   !    Conjugate Residual Method Inner Iteration
2324   !**************************************************
2325   SUBROUTINE CR_INNER (MD,NNZERO,INNER_ITER,AA_AUX,JA_AUX,IA_AUX,X_OLD,B,
              X)
2326   !-------------------------------------------------
2327   USE PARAM, ONLY                      : TOL
2328   IMPLICIT NONE
2329   INTEGER                              :: TL,MD,NNZERO,IDIAG,IERR,
              INNER_ITER,I
2330   REAL(8),DIMENSION(MD)                :: X_OLD,X,P_OLD,P,R_OLD,R,AP,AP_OLD
              ,AR_OLD,AR,B
2331   INTEGER,DIMENSION(MD+1)               :: IA_AUX,IAR
2332   INTEGER,DIMENSION(2*MD-1)             :: IND
2333   REAL(8)                              :: ALPHA,BETA
2334   REAL(8),DIMENSION(NNZERO)             :: AA_AUX,AAR
2335   INTEGER,DIMENSION(NNZERO)             :: JA_AUX,JAR
2336   REAL(8),ALLOCATABLE                   :: SA(:),DIAG(:,:),COEF(:,:)
2337   INTEGER,ALLOCATABLE                   :: IJA(:),IOFF(:),JCOEF(:,:)
2338   !-------------------------------------------------
2339   CALL TOTAL_LENGTH         (MD,IA_AUX,TL)
2340   ALLOCATE                  (SA(TL),IJA(TL))
2341   CALL CSRMSR               (TL,MD,NNZERO,AA_AUX,JA_AUX,IA_AUX,SA,IJA)
2342   CALL INFDIA               (MD,NNZERO,JA_AUX,IA_AUX,IND,IDIAG)
2343   ALLOCATE                  (COEF(MD,IDIAG),JCOEF(MD,IDIAG))
2344   ALLOCATE                  (IOFF(IDIAG),DIAG(MD,IDIAG))
2345   CALL CSRELL               (MD,NNZERO,AA_AUX,JA_AUX,IA_AUX,MD,COEF,
              JCOEF,MD,IDIAG,IERR)
2346   CALL CSRDIA               (MD,NNZERO,IDIAG,10,AA_AUX,JA_AUX,IA_AUX,MD,
              DIAG,IOFF,AAR,JAR,IAR,IND)
2347   !-------------------------------------------------
2348   CALL INITIAL_RESIDUAL (MD,NNZERO,AA_AUX,JA_AUX,IA_AUX,B,X_OLD,R_OLD)
2349   !-------------------------------------------------
2350   P_OLD=R_OLD
2351   CALL MSR_MAT_V_PRODUCT (TL,MD,SA,IJA,P_OLD,AP_OLD)
2352   DO I=1,INNER_ITER
2353   !-------------------------------------------------
2354   ! Stores the matrix in compressed spars row and multiplies by
2355   ! vector
2356   !-------------------------------------------------
2357   !CALL CSR_MAT_V_PRODUCT     (MD,NNZERO,AA_AUX,JA_AUX,IA_AUX,R_OLD,
              AR_OLD)
2358   !-------------------------------------------------
2359   ! Stores the matrix in modified spars row and multiplies by
2360   ! vector
2361   !-------------------------------------------------
2362      CALL MSR_MAT_V_PRODUCT (TL,MD,SA,IJA,R_OLD,AR_OLD)
2363   !-------------------------------------------------
2364   ! Stores the matrix in Ellpack/Itpack format and multiplies by
2365   ! vector
2366   !-------------------------------------------------
```

```
2367  !CALL ELLPACK_ITPACK_MAT_V_PRODUCT (MD,R_OLD,AR_OLD,MD,IDIAG,COEF,JCOEF
      )
2368  !------------------------------------------------
2369  ! Stores the matrix in Diagonal format and multiplies by
2370  ! vector
2371  !------------------------------------------------
2372  !CALL DIAGONAL_MAT_V_PRODUCT (MD,R_OLD,AR_OLD,DIAG,MD,IDIAG,IOFF)
2373  !------------------------------------------------
2374     ALPHA=DOT_PRODUCT(R_OLD,AR_OLD)/DOT_PRODUCT(AP_OLD,AP_OLD)
2375     X=X_OLD+ALPHA*P_OLD
2376     R=R_OLD-ALPHA*AP_OLD
2377  !------------------------------------------------
2378  ! Stores the matrix in compressed spars row and multiplies by
2379  ! vector
2380  !------------------------------------------------
2381  !CALL CSR_MAT_V_PRODUCT      (MD,NNZERO,AA_AUX,JA_AUX,IA_AUX,R,AR)
2382  !------------------------------------------------
2383  ! Stores the matrix in modified spars row and multiplies by
2384  ! vector
2385  !------------------------------------------------
2386     CALL MSR_MAT_V_PRODUCT (TL,MD,SA,IJA,R,AR)
2387  !------------------------------------------------
2388  ! Stores the matrix in Ellpack/Itpack format and multiplies by
2389  ! vector
2390  !------------------------------------------------
2391  !CALL ELLPACK_ITPACK_MAT_V_PRODUCT (MD,R,AR,MD,IDIAG,COEF,JCOEF)
2392  !------------------------------------------------
2393  ! Stores the matrix in Diagonal format and multiplies by
2394  ! vector
2395  !------------------------------------------------
2396  !CALL DIAGONAL_MAT_V_PRODUCT (MD,R,AR,DIAG,MD,IDIAG,IOFF)
2397  !------------------------------------------------
2398     BETA=DOT_PRODUCT(R,AR)/DOT_PRODUCT(R_OLD,AR_OLD)
2399     P=R+BETA*P_OLD
2400     AP=AR+BETA*AP_OLD
2401     P_OLD=P
2402     X_OLD=X
2403     R_OLD=R
2404     AP_OLD=AP
2405  END DO
2406  !------------------------------------------------
2407  END SUBROUTINE CR_INNER
2408  !*************************************************
2409  !                GMRES* Method
2410  !*************************************************
2411  SUBROUTINE GMRES_STAR (MD,NNZERO,AA_AUX,JA_AUX,IA_AUX,XX_OLD,R_OLD,X)
2412  !------------------------------------------------
2413  USE PARAM, ONLY              : TOL,ITMAX
2414  IMPLICIT NONE
2415  INTEGER                      :: I,K,MD,NNZERO,IERR,TL,IDIAG
2416  REAL(8)                      :: NORMC,NORM,ALPHA,CR,S,SR
2417  REAL(8),ALLOCATABLE          :: SA(:),DIAG(:,:),COEF(:,:)
2418  INTEGER,ALLOCATABLE          :: IJA(:),IOFF(:),JCOEF(:,:)
2419  REAL(8),DIMENSION(NNZERO)    :: AA_AUX,AAR
2420  INTEGER,DIMENSION(NNZERO)    :: JA_AUX,JAR
2421  INTEGER,DIMENSION(MD+1)      :: IA_AUX,IAR
2422  INTEGER,DIMENSION(2*MD-1)    :: IND
2423  REAL(8),DIMENSION(MD)        :: ZM,X_OLD,XX_OLD,R_OLD,X,R,C
2424  REAL(8),DIMENSION(ITMAX,MD)  :: CC,U
2425  !------------------------------------------------
2426  CALL TOTAL_LENGTH           (MD,IA_AUX,TL)
2427  ALLOCATE                    (SA(TL),IJA(TL))
2428  CALL CSRMSR                 (TL,MD,NNZERO,AA_AUX,JA_AUX,IA_AUX,SA,IJA)
2429  CALL INFDIA                 (MD,NNZERO,JA_AUX,IA_AUX,IND,IDIAG)
2430  ALLOCATE                    (COEF(MD,IDIAG),JCOEF(MD,IDIAG))
2431  ALLOCATE                    (IOFF(IDIAG),DIAG(MD,IDIAG))
```

```
2432 CALL CSRELL                      (MD,NNZERO,AA_AUX,JA_AUX,IA_AUX,MD,COEF,
         JCOEF,MD,IDIAG,IERR)
2433 CALL CSRDIA                      (MD,NNZERO,IDIAG,10,AA_AUX,JA_AUX,IA_AUX,MD,
         DIAG,IOFF,AAR,JAR,IAR,IND)
2434 !----------------------------------------------------
2435 DO I=1,ITMAX
2436 !----------------------------------------------------
2437    CALL CG_INNER                 (MD,NNZERO,20,AA_AUX,JA_AUX,IA_AUX,XX_OLD,
         R_OLD,ZM)
2438    !CALL ILU0_PCG_INNER          (MD,NNZERO,20,AA_AUX,JA_AUX,IA_AUX,XX_OLD,
         R_OLD,ZM)
2439    !CALL BCGSTAB_INNER           (MD,NNZERO,20,AA_AUX,JA_AUX,IA_AUX,XX_OLD,
         R_OLD,ZM)
2440    !CALL ILU0_PBCGSTAB_INNER     (MD,NNZERO,20,AA_AUX,JA_AUX,IA_AUX,XX_OLD,
         R_OLD,ZM)
2441    !CALL CGS_INNER               (MD,NNZERO,20,AA_AUX,JA_AUX,IA_AUX,XX_OLD,
         R_OLD,ZM)
2442    !CALL ILU0_PCGS_INNER         (MD,NNZERO,20,AA_AUX,JA_AUX,IA_AUX,XX_OLD,
         R_OLD,ZM)
2443    !CALL CR_INNER                (MD,NNZERO,20,AA_AUX,JA_AUX,IA_AUX,XX_OLD,
         R_OLD,ZM)
2444 !----------------------------------------------------
2445 ! Stores the matrix in compressed spars row and multiplies by
2446 ! vector
2447 !----------------------------------------------------
2448 !CALL CSR_MAT_V_PRODUCT    (MD,NNZERO,AA_AUX,JA_AUX,IA_AUX,ZM,C)
2449 !----------------------------------------------------
2450 ! Stores the matrix in modified spars row and multiplies by
2451 ! vector
2452 !----------------------------------------------------
2453 !CALL MSR_MAT_V_PRODUCT (TL,MD,SA,IJA,ZM,C)
2454 !----------------------------------------------------
2455 ! Stores the matrix in Ellpack/Itpack format and multiplies by
2456 ! vector
2457 !----------------------------------------------------
2458 !CALL ELLPACK_ITPACK_MAT_V_PRODUCT (MD,ZM,C,MD,IDIAG,COEF,JCOEF)
2459 !----------------------------------------------------
2460 ! Stores the matrix in Diagonal format and multiplies by
2461 ! vector
2462 !----------------------------------------------------
2463    CALL DIAGONAL_MAT_V_PRODUCT (MD,ZM,C,DIAG,MD,IDIAG,IOFF)
2464 !----------------------------------------------------
2465    DO K=1,I-1
2466       ALPHA=DOT_PRODUCT(CC(K,:),C)
2467       C=C-ALPHA*CC(K,:)
2468       ZM=ZM-ALPHA*U(K,:)
2469    END DO
2470    S=DOT_PRODUCT(C,C)
2471    NORMC=DSQRT(S)
2472    CC(I,:)=C/NORMC
2473    U(I,:)=ZM/NORMC
2474    CR=DOT_PRODUCT(CC(I,:),R_OLD)
2475    X=X_OLD+CR*U(I,:)
2476    R=R_OLD-CR*CC(I,:)
2477    SR=DOT_PRODUCT(R,R)
2478    NORM=DSQRT(SR)/MD
2479    R_OLD=R
2480    X_OLD=X
2481    !PRINT*,NORM
2482    !CALL SUCCESSIVE_SOL (NORM,I)
2483    IF (NORM.LT.TOL) THEN
2484       EXIT
2485    ELSE IF (I.EQ.ITMAX.AND.NORM.GT.TOL) THEN
2486       PRINT*, 'convergency is not met by this number of iteration'
2487    END IF
2488 END DO
2489 !----------------------------------------------------
```

```
2490 END SUBROUTINE GMRES_STAR
2491 !**************************************************
2492 !        Conjugate Residual Subroutine
2493 !**************************************************
2494 SUBROUTINE CR (MD,NNZERO,AA_AUX,JA_AUX,IA_AUX,X_OLD,R_OLD,X)
2495 !----------------------------------------------
2496 USE PARAM, ONLY                  : TOL
2497 IMPLICIT NONE
2498 INTEGER                          :: ITER,TL,MD,NNZERO,IDIAG,IERR
2499 REAL(8),DIMENSION(MD)            :: X_OLD,X,P_OLD,P,R_OLD,R,AP,AP_OLD
       ,AR_OLD,AR
2500 INTEGER,DIMENSION(MD+1)          :: IA_AUX,IAR
2501 INTEGER,DIMENSION(2*MD-1)        :: IND
2502 REAL(8)                          :: NORM,S,ALPHA,BETA
2503 REAL(8),DIMENSION(NNZERO)        :: AA_AUX,AAR
2504 INTEGER,DIMENSION(NNZERO)        :: JA_AUX,JAR
2505 REAL(8),ALLOCATABLE              :: SA(:),DIAG(:,:),COEF(:,:)
2506 INTEGER,ALLOCATABLE              :: IJA(:),IOFF(:),JCOEF(:,:)
2507 !----------------------------------------------
2508 CALL TOTAL_LENGTH               (MD,IA_AUX,TL)
2509 ALLOCATE                        (SA(TL),IJA(TL))
2510 CALL CSRMSR                     (TL,MD,NNZERO,AA_AUX,JA_AUX,IA_AUX,SA,IJA)
2511 CALL INFDIA                     (MD,NNZERO,JA_AUX,IA_AUX,IND,IDIAG)
2512 ALLOCATE                        (COEF(MD,IDIAG),JCOEF(MD,IDIAG))
2513 ALLOCATE                        (IOFF(IDIAG),DIAG(MD,IDIAG))
2514 CALL CSRELL                     (MD,NNZERO,AA_AUX,JA_AUX,IA_AUX,MD,COEF,
       JCOEF,MD,IDIAG,IERR)
2515 CALL CSRDIA                     (MD,NNZERO,IDIAG,10,AA_AUX,JA_AUX,IA_AUX,MD,
       DIAG,IOFF,AAR,JAR,IAR,IND)
2516 !----------------------------------------------
2517 P_OLD=R_OLD
2518 CALL MSR_MAT_V_PRODUCT (TL,MD,SA,IJA,P_OLD,AP_OLD)
2519 NORM=1.D0
2520 ITER=0
2521 DO WHILE (NORM.GT.TOL)
2522    ITER=ITER+1
2523    !PRINT*,ITER
2524 !----------------------------------------------
2525 ! Stores the matrix in compressed spars row and multiplies by
2526 ! vector
2527 !----------------------------------------------
2528    !CALL CSR_MAT_V_PRODUCT    (MD,NNZERO,AA_AUX,JA_AUX,IA_AUX,R_OLD,
       AR_OLD)
2529 !----------------------------------------------
2530 ! Stores the matrix in modified spars row and multiplies by
2531 ! vector
2532 !----------------------------------------------
2533    !CALL MSR_MAT_V_PRODUCT (TL,MD,SA,IJA,R_OLD,AR_OLD)
2534 !----------------------------------------------
2535 ! Stores the matrix in Ellpack/Itpack format and multiplies by
2536 ! vector
2537 !----------------------------------------------
2538    !CALL ELLPACK_ITPACK_MAT_V_PRODUCT (MD,R_OLD,AR_OLD,MD,IDIAG,COEF,
       JCOEF)
2539 !----------------------------------------------
2540 ! Stores the matrix in Diagonal format and multiplies by
2541 ! vector
2542 !----------------------------------------------
2543    CALL DIAGONAL_MAT_V_PRODUCT (MD,R_OLD,AR_OLD,DIAG,MD,IDIAG,IOFF)
2544 !----------------------------------------------
2545    ALPHA=DOT_PRODUCT(R_OLD,AR_OLD)/DOT_PRODUCT(AP_OLD,AP_OLD)
2546    X=X_OLD+ALPHA*P_OLD
2547    R=R_OLD-ALPHA*AP_OLD
2548 !----------------------------------------------
2549 ! Stores the matrix in compressed spars row and multiplies by
2550 ! vector
2551 !----------------------------------------------
```

```
2552    !CALL CSR_MAT_V_PRODUCT     (MD,NNZERO,AA_AUX,JA_AUX,IA_AUX,R,AR)
2553  !-----------------------------------------------
2554  ! Stores the matrix in modified spars row and multiplies by
2555  ! vector
2556  !-----------------------------------------------
2557    !CALL MSR_MAT_V_PRODUCT (TL,MD,SA,IJA,R,AR)
2558  !-----------------------------------------------
2559  ! Stores the matrix in Ellpack/Itpack format and multiplies by
2560  ! vector
2561  !-----------------------------------------------
2562    !CALL ELLPACK_ITPACK_MAT_V_PRODUCT (MD,R,AR,MD,IDIAG,COEF,JCOEF)
2563  !-----------------------------------------------
2564  ! Stores the matrix in Diagonal format and multiplies by
2565  ! vector
2566  !-----------------------------------------------
2567    CALL DIAGONAL_MAT_V_PRODUCT (MD,R,AR,DIAG,MD,IDIAG,IOFF)
2568  !-----------------------------------------------
2569    BETA=DOT_PRODUCT(R,AR)/DOT_PRODUCT(R_OLD,AR_OLD)
2570    P=R+BETA*P_OLD
2571    AP=AR+BETA*AP_OLD
2572    S=DOT_PRODUCT(R,R)
2573    NORM=DSQRT(S)/MD
2574  !-----------------------------------------------
2575    P_OLD=P
2576    X_OLD=X
2577    R_OLD=R
2578    AP_OLD=AP
2579    !PRINT*,NORM
2580    !CALL SUCCESSIVE_SOL (NORM,ITER)
2581  END DO
2582  !-----------------------------------------------
2583  END SUBROUTINE CR
```

Bibliography

RR Akhunov, SP Kuksenko, VK Salov, and TR Gazizov. Optimization of the ilu (0) factorization algorithm with the use of compressed sparse row format. *Journal of Mathematical Sciences*, 191(1):19–27, 2013.

Walter Edwin Arnoldi. The principle of minimized iterations in the solution of the matrix eigenvalue problem. *Quarterly of applied mathematics*, 9(1):17–29, 1951.

Owe Axelsson and Vincent Allan Barker. *Finite element solution of boundary value problems: theory and computation*, volume 35. Siam, 1984.

Owe Axelsson and Panayot S Vassilevski. A black box generalized conjugate gradient solver with inner iterations and variable-step preconditioning. *SIAM Journal on Matrix Analysis and Applications*, 12(4):625–644, 1991.

Zhaojun Bai, James Demmel, Jack Dongarra, Axel Ruhe, and Henk van der Vorst. *Templates for the solution of algebraic eigenvalue problems: a practical guide*. SIAM, 2000.

Richard Barrett, Michael W Berry, Tony F Chan, James Demmel, June Donato, Jack Dongarra, Victor Eijkhout, Roldan Pozo, Charles Romine, and Henk Van der Vorst. *Templates for the solution of linear systems: building blocks for iterative methods*, volume 43. Siam, 1994.

Ronald F Boisvert, Roldan Pozo, Karin Remington, Richard F Barrett, and Jack J Dongarra. Matrix market: a web resource for test matrix collections. In *Quality of Numerical Software*, pages 125–137. Springer, 1997.

Tony F Chan, Efstratios Gallopoulos, Valeria Simoncini, Tedd Szeto, and Charles H Tong. A quasi-minimal residual variant of the bi-cgstab algorithm for nonsymmetric systems. *SIAM Journal on Scientific Computing*, 15(2): 338–347, 1994.

Ke Chen. *Matrix preconditioning techniques and applications*, volume 19. Cambridge University Press, 2005.

Xinhai Chen, Peizhen Xie, Lihua Chi, Jie Liu, and Chunye Gong. An efficient simd compression format for sparse matrix-vector multiplication. *Concurrency and Computation: Practice and Experience*, 30(23):e4800, 2018.

Alexandre Joel Chorin. Numerical solution of incompressible flow problems. *Studies in Numerical Analysis*, 2:64–71, 1968.

Ian S Duff. User's guide for the harwell-boeing sparse matrix collection. *Technical Report, Rutherford Appleton Laboratory*, 1992.

Vance Faber and Thomas Manteuffel. Necessary and sufficient conditions for the existence of a conjugate gradient method. *SIAM Journal on Numerical Analysis*, 21(2):352–362, 1984.

Joel H Ferziger. Simulation of incompressible turbulent flows. *Journal of Computational Physics*, 69:1–48, 1987.

Roger Fletcher. Conjugate gradient methods for indefinite systems. In *Numerical analysis*, pages 73–89. Springer, 1976.

Roland W Freund. A transpose-free quasi-minimal residual algorithm for non-hermitian linear systems. *SIAM journal on scientific computing*, 14(2): 470–482, 1993.

Roland W Freund, Gene H Golub, and Noël M Nachtigal. Iterative solution of linear systems. *Acta numerica*, 1:57–100, 1992.

Gene H Golub and Charles F Van Loan. *Matrix computations*. The Johns Hopkins University Press, Baltimore, USA, 1989.

Martin H Gutknecht. Variants of bicgstab for matrices with complex spectrum. *SIAM journal on scientific computing*, 14(5):1020–1033, 1993.

Gundolf Haase, Manfred Liebmann, and Gernot Plank. A hilbert-order multiplication scheme for unstructured sparse matrices. *International Journal of Parallel, Emergent and Distributed Systems*, 22(4):213–220, 2007.

L Hageman and M Young. Applied iterative methods academic. *New York*, 1981.

Magnus Rudolph Hestenes and Eduard Stiefel. *Methods of conjugate gradients for solving linear systems*, volume 49. NBS Washington, DC, 1952.

Joe D Hoffman. *Numerical methods for engineers and scientists*, second editionrevised and expanded, by marcel dekker. *Inc.* New York, 2001.

Eun-Jin Im and Katherine Yelick. Optimizing sparse matrix computations for register reuse in sparsity. In *International Conference on Computational Science*, pages 127–136. Springer, 2001.

Nikolaj Nikolaevič Janenko. *The method of fractional steps*. Springer, 1971.

Olin G Johnson, Charles A Micchelli, and George Paul. Polynomial preconditioners for conjugate gradient calculations. *SIAM Journal on Numerical Analysis*, 20(2):362–376, 1983.

Shmuel Kaniel. Estimates for some computational techniques in linear algebra. *Mathematics of Computation*, 20(95):369–378, 1966.

John Kim and Parviz Moin. Application of a fractional-step method to incompressible navier-stokes equations. *Journal of computational physics*, 59 (2):308–323, 1985.

David R Kincaid and David M Young. The itpack project: Past, present, and future. In *Elliptic Problem Solvers*, pages 53–63. Elsevier, 1984.

Ming-Chih Lai and Charles S Peskin. An immersed boundary method with formal second-order accuracy and reduced numerical viscosity. *Journal of computational Physics*, 160(2):705–719, 2000.

Cornelius Lanczos. *An iteration method for the solution of the eigenvalue problem of linear differential and integral operators*. United States Governm. Press Office Los Angeles, CA, 1950.

Cornelius Lanczos. Solution of systems of linear equations by minimized iterations. *J. Res. Nat. Bur. Standards*, 49(1):33–53, 1952.

Chuan-Chieh Liao, Yu-Wei Chang, Chao-An Lin, and JM McDonough. Simulating flows with moving rigid boundary using immersed-boundary method. *Computers & Fluids*, 39(1):152–167, 2010.

James M McDonough. Lectures in computational fluid dynamics of incompressible flow: Mathematics, algorithms and implementations. 2007.

James M McDonough. Lectures on computational numerical analysis of partial differential equations. 2008.

Gérard Meurant. *The Lanczos and conjugate gradient algorithms: from theory to finite precision computations*, volume 19. SIAM, 2006.

Carl D Meyer. *Matrix analysis and applied linear algebra*, volume 71. Siam, 2000.

Eurıpides Montagne and Anand Ekambaram. An optimal storage format for sparse matrices. *Information Processing Letters*, 90(2):87–92, 2004.

N Nachtigal, Lothar Reichel, and L Trefethen. A hybrid gmres algorithm for nonsymmetric matrix iterations. Technical report, Tech. Rep. 90-7, MIT, Cambridge, MA, 1990.

Noël M Nachtigal, Satish C Reddy, and Lloyd N Trefethen. How fast are nonsymmetric matrix iterations? *SIAM Journal on Matrix Analysis and Applications*, 13(3):778–795, 1992.

Ali Pinar and Michael T Heath. Improving performance of sparse matrix-vector multiplication. In *SC'99: Proceedings of the 1999 ACM/IEEE Conference on Supercomputing*, pages 30–30. IEEE, 1999.

Ugo Piomelli and Elias Balaras. Wall-layer models for large-eddy simulations. *Annual review of fluid mechanics*, 34(1):349–374, 2002.

Sergio Pissanetzky. *Sparse Matrix Technology-electronic edition*. Academic Press, 1984.

C Pozrikidis. On the relationship between the pressure and the projection function in the numerical computation of viscous incompressible flow. *European Journal of Mechanics-B/Fluids*, 22(2):105–121, 2003.

William H Press, Saul A Teukolsky, William T Vetterling, and Brian P Flannery. *Numerical recipes in Fortran 77: the art of scientific computing*, volume 2. Cambridge university press Cambridge, 1992.

John K Reid. On the method of conjugate gradients for the solution of large sparse systems of linear equations. In *Pro. the Oxford conference of institute of mathmatics and its applications*, pages 231–254, 1971.

Dietmar Rempfer. On boundary conditions for incompressible navier-stokes problems. *Applied Mechanics Reviews*, 59(3):107–125, 2006.

Anatol Roshko. On the wake and drag of bluff bodies. *Journal of the aeronautical sciences*, 22(2):124–132, 1955.

Youcef Saad. Practical use of polynomial preconditionings for the conjugate gradient method. *SIAM Journal on Scientific and Statistical Computing*, 6 (4):865–881, 1985.

Youcef Saad. A flexible inner-outer preconditioned gmres algorithm. *SIAM Journal on Scientific Computing*, 14(2):461–469, 1993.

Youcef Saad and Martin H Schultz. Gmres: A generalized minimal residual algorithm for solving nonsymmetric linear systems. *SIAM Journal on scientific and statistical computing*, 7(3):856–869, 1986.

Yousef Saad. Iterative methods for sparse linear systems pws. New York, 1996.

Yousef Saad. *Numerical methods for large eigenvalue problems: revised edition*, volume 66. Siam, 2011.

Yousef Saad and Henk A Van Der Vorst. Iterative solution of linear systems in the 20th century. *Journal of Computational and Applied Mathematics*, 123 (1-2):1–33, 2000.

Gemma Sanjuan, Tomàs Margalef, and Ana Cortés. Hybrid application to accelerate wind field calculation. *Journal of Computational Science*, 17: 576–590, 2016.

Rukhsana Shahnaz and Anila Usman. An efficient sparse matrix vector multiplication on distributed memory parallel computers. *International Journal of Computer Science and Network Security*, 7(1):77–82, 2007.

Rukhsana Shahnaz and Anila Usman. Blocked-based sparse matrix-vector multiplication on distributed memory parallel computers. *Int. Arab J. Inf. Technol.*, 8(2):130–136, 2011.

ALF Lima E Silva, A Silveira-Neto, and JJR Damasceno. Numerical simulation of two-dimensional flows over a circular cylinder using the immersed boundary method. *Journal of Computational Physics*, 189(2):351–370, 2003.

Gerard LG Sleijpen and Diederik R Fokkema. Bicgstab (l) for linear equations involving unsymmetric matrices with complex spectrum. *Electronic Transactions on Numerical Analysis*, 1(11):2000, 1993.

FS Smailbegovic, Georgi N Gaydadjiev, and Stamatis Vassiliadis. Sparse matrix storage format. In *Proceedings of the 16th Annual Workshop on Circuits, Systems and Signal Processing, ProRisc 2005*, pages 445–448, 2005.

Peter Sonneveld. Cgs, a fast lanczos-type solver for nonsymmetric linear systems. *SIAM journal on scientific and statistical computing*, 10(1):36–52, 1989.

Roger Temam. Sur l'approximation de la solution des équations de navier-stokes par la méthode des pas fractionnaires (ii). *Archive for Rational Mechanics and Analysis*, 33(5):377–385, 1969.

Sivan Toledo. Improving the memory-system performance of sparse-matrix vector multiplication. *IBM Journal of research and development*, 41(6): 711–725, 1997.

Charles H Tong. A comparative study of preconditioned lanczos methods for nonsymmetric linear system. Technical report, Sandia National Labs., Livermore, CA (United States), 1992.

Abraham Van der Sluis and Henk A van der Vorst. The rate of convergence of conjugate gradients. *Numerische Mathematik*, 48(5):543–560, 1986.

Henk A Van der Vorst. Bi-cgstab: A fast and smoothly converging variant of bi-cg for the solution of nonsymmetric linear systems. *SIAM Journal on scientific and Statistical Computing*, 13(2):631–644, 1992.

Henk A Van Der Vorst. Efficient and reliable iterative methods for linear systems. *Journal of Computational and Applied Mathematics*, 149(1):251–265, 2002.

Henk A Van der Vorst. *Iterative Krylov methods for large linear systems*, volume 13. Cambridge University Press, 2003.

Henk A Van der Vorst and Cornelis Vuik. Gmresr: a family of nested gmres methods. *Numerical linear algebra with applications*, 1(4):369–386, 1994.

Valentin Vassilevich Voevodin. The problem of nonselfadjoint extension of the conjugate gradient method is closed. *Zhurnal Vychislitel'noi Matematiki i Matematicheskoi Fiziki*, 23(2):477–479, 1983.

Hendrik Albertus Vorst and Cornelis Vuik. *GMRESR: A family of nested GMRES methods*. TU, Faculteit der Technische Wiskunde en Informatica, 1991.

Richard W Vuduc and Hyun-Jin Moon. Fast sparse matrix-vector multiplication by exploiting variable block structure. In *International Conference on High Performance Computing and Communications*, pages 807–816. Springer, 2005.

Charles HK Williamson. Vortex dynamics in the cylinder wake. *Annual review of fluid mechanics*, 28(1):477–539, 1996.

U Meier Yang. *Preconditioned conjugate gradient-like methods for nonsymmetric linear systems*. University of Illinois at Urbana-Champaign. Center for Supercomputing Center for Supercomputing Research and Development [CSRD], 1992.

AN Yzelman and Rob H Bisseling. Cache-oblivious sparse matrix–vector multiplication by using sparse matrix partitioning methods. *SIAM Journal on Scientific Computing*, 31(4):3128–3154, 2009.

Index

Krylov Subspace Methods with Application in Incompressible Fluid Flow Solvers, First Edition.
Iman Farahbakhsh.
© 2020 John Wiley & Sons Ltd. Published 2020 by John Wiley & Sons Ltd.
Companion Website: www.wiley.com/go/Farahbakhs/KrylovSubspaceMethods